Zongmin Ma (Ed.)

Soft Computing in Ontologies and Semantic Web

Studies in Fuzziness and Soft Computing, Volume 204

Editor-in-chief
Prof. Janusz Kacprzyk
Systems Research Institute
Polish Academy of Sciences
ul. Newelska 6
01-447 Warsaw
Poland
E-mail: kacprzyk@ibspan.waw.pl

Further volumes of this series
can be found on our homepage:
springer.com

Vol. 188. James J. Buckley, Leonard J. Jowers
Simulating Continuous Fuzzy Systems, 2006
ISBN 3-540-28455-9

Vol. 189. Hans Bandemer
Mathematics of Uncertainty, 2006
ISBN 3-540-28457-5

Vol. 190. Ying-ping Chen
Extending the Scalability of Linkage Learning Genetic Algorithms, 2006
ISBN 3-540-28459-1

Vol. 191. Martin V. Butz
Rule-Based Evolutionary Online Learning Systems, 2006
ISBN 3-540-25379-3

Vol. 192. Jose A. Lozano, Pedro Larrañaga, Iñaki Inza, Endika Bengoetxea (Eds.)
Towards a New Evolutionary Computation, 2006
ISBN 3-540-29006-0

Vol. 193. Ingo Glöckner
Fuzzy Quantifiers: A Computational Theory, 2006
ISBN 3-540-29634-4

Vol. 194. Dawn E. Holmes, Lakhmi C. Jain (Eds.)
Innovations in Machine Learning, 2006
ISBN 3-540-30609-9

Vol. 195. Zongmin Ma
Fuzzy Database Modeling of Imprecise and Uncertain Engineering Information, 2006
ISBN 3-540-30675-7

Vol. 196. James J. Buckley
Fuzzy Probability and Statistics, 2006
ISBN 3-540-30841-5

Vol. 197. Enrique Herrera-Viedma, Gabriella Pasi, Fabio Crestani (Eds.)
Soft Computing in Web Information Retrieval, 2006
ISBN 3-540-31588-8

Vol. 198. Hung T. Nguyen, Berlin Wu
Fundamentals of Statistics with Fuzzy Data, 2006
ISBN 3-540-31695-7

Vol. 199. Zhong Li
Fuzzy Chaotic Systems, 2006
ISBN 3-540-33220-0

Vol. 200. Kai Michels, Frank Klawonn, Rudolf Kruse, Andreas Nürnberger
Fuzzy Control, 2006
ISBN 3-540-31765-1

Vol. 201. Cengiz Kahraman (Ed.)
Fuzzy Applications in Industrial Engineering, 2006
ISBN 3-540-33516-1

Vol. 202. Patrick Doherty, Witold Łukaszewicz, Andrzej Skowron, Andrzej Szałas
Knowledge Representation Techniques: A Rough Set Approach, 2006
ISBN 3-540-33518-8

Vol. 203. Gloria Bordogna, Giuseppe Psaila (Eds.)
Flexible Databases Supporting Imprecision and Uncertainty, 2006
ISBN 3-540-33288-X

Vol. 204. Zongmin Ma (Ed.)
Soft Computing in Ontologies and Semantic Web, 2006
ISBN 3-540-33472-6

Zongmin Ma
(Ed.)

Soft Computing in Ontologies and Semantic Web

Springer

Dr. Zongmin Ma
College of Information Science and Engineering
Northeastern University
Shenyang, Liaoning 110004, China
E-mail: zongmin_ma@yahoo.com

ISSN print edition: 1434-9922
ISSN electronic edition: 1860-0808

ISBN 978-3-642-07010-5 e-ISBN 978-3-540-33473-6

Springer is a part of Springer Science+Business Media
springer.com

Softcover reprint of the hardcover 1st edition 2006

Cover design: Erich Kirchner, Heidelberg

Preface

Being the important means by which people acquire and publish information nowadays, the Web has been a huge information resource depository all around the world and the huge amount of information on the Web is getting larger and larger every day. It is becoming very crucial for computer programs to deal with information on the Web automatically and intelligently. But most of today's Web content is suitable for human consumption. The Semantic Web is a vision that has sparked a wide-ranging enthusiasm for a generation of the Web. The Semantic Web has emerged as an extension of the current Web in which information is given well-defined meaning, enabling computers and people to better work in cooperation. The central idea of the Semantic Web is to make the Web more understandable to computer programs so that people can make more use of this gigantic asset.

The Semantic Web is generally built on syntaxes which use URIs to represent data, usually in triple based structures: i.e. many triples of URI data that can be held in databases, or interchanged in the World Wide Web using a set of particular syntaxes developed especially for the task. These syntaxes are called Resource Description Framework (RDF) syntaxes. The layer above the syntax is the simple datatyping model. The RDF Schema is designed to be a simple datatyping model for the RDF. The Web Ontology Language (OWL) is a language as an ontology language based upon the RDF. OWL takes the RDF Schema a step further, by giving us more in-depth properties and classes. The next step in the architecture of the Semantic Web is trust and proof.

In the real world, human knowledge and natural language have a big deal of imprecision and vagueness. While the Semantic Web concept and research attracts attention, as long as there will be used two-valued-based logical methods no progress will be expected in handling ill-structured, uncertain or imprecise information encountered in the real world knowledge. Fuzzy logic, probability, and more generally soft computing, have been applied in a large number and a wide variety of applications, with a real-world impact across a wide array of domains with human-like behavior and reasoning. Soft computing has been a crucial means of implementing machine intelligence. Therefore, soft computing cannot be ignored in order to bridge the gap between human-understandable soft logic and machine-readable hard logic. None of the usual logical requirements can be guaranteed: there is no centrally defined format for data, no guarantee of truth for assertions made, and no guarantee of consistency. It can be believed that soft computing can play an important and positive role in the development of the Semantic Web. It should be noticed, however, that soft computing may not be assumed to be the (only) basis for the Semantic Web, but its related concepts and techniques will certainly reinforce the systems classically developed within the W3C. Currently the research and development of soft computing in the area of ontologies and Semantic Web are attracting an increased attention.

This book covers in a great depth the fast growing topic of tools, techniques and applications of soft computing (e.g., fuzzy logic, genetic algorithms, neural networks, rough sets, Bayesian networks, and other probabilistic techniques) in the ontologies and Semantic Web. It is shown how components of the Semantic Web (like the RDF, Description Logics, ontologies) can be covered with a soft computing focus. This book aims to provide a single account of current studies in soft computing approaches to the ontologies and the Semantic Web. The objective of the book is to provide the state of the art information to researchers, practitioners, and graduate students of the Web intelligence, and at the same time serving the information technology professional faced with non-traditional applications that make the application of conventional approaches difficult or impossible.

This book, which consists of ten chapters, is organized into two major sections. The first section discusses the probability in the ontologies and Semantic Web; it consists of the first four chapters. The next six chapters, covering fuzzy logic in the ontologies and the Semantic Web, comprise the second section.

Chapter 1 describes an on-going research on developing a probabilistic framework for the modeling of uncertainty in the Semantic Web ontologies based on Bayesian networks. New OWL classes, which can be used to encode probability constraints for ontology classes and relations in OWL, and a set of rules for translating OWL ontology taxonomy into Bayesian network DAG, are defined. Also a new algorithm, D-IPFP, for an efficient construction of the CPTs is provided.

Chapter 2 presents a probabilistic approach to the problem that the Semantic Web ontologies based on crisp logic do not provide well-defined means for expressing uncertainty. In the method, degrees of subsumption, i.e., overlap between concepts, can be modeled and computed efficiently using Bayesian networks based on RDF(S) ontologies. Degrees of overlap indicate how well an individual data item matches a query concept which can be used as a well-defined measure of relevance in information retrieval tasks.

Chapter 3 introduces oPLMap, a formal framework for automatically learning mapping rules between heterogeneous Web directories, a crucial step towards integrating the ontologies and their instances in the Semantic Web.

The approach is based on the Horn predicate logics and probability theory which allows for dealing with uncertain mappings (for cases where there is no exact correspondence between classes), and can be extended towards complex ontology models. Different components are combined for finding suitable mapping candidates (together with their weights), and the set of rules with maximum matching probability is selected.

Chapter 4 describes an approach to the representation and processing of knowledge, based on the SP theory of computing and cognition. The benefits of the SP approach are the simplicity and comprehensibility in the representation of knowledge, an ability to cope with errors and uncertainties in knowledge, and capabilities for 'intelligent' processing of knowledge, including probabilistic reasoning, pattern recognition, information retrieval, unsupervised learning, planning and problem solving. The approach proposed has strengths that complement others such as those currently proposed for the Semantic Web.

Chapter 5 discusses problems related to the design of ambient intelligence (AmI) environments and demonstrates how it is possible to obtain the autonomy, independence and distribution of computational resources by means of a hybrid approach. The autonomy is realized using the well-know fuzzy logic theory which permits to control, in an automatic way, different devices composing the AmI framework; the XML technologies allow to realize the independence feature; the Web service technologies allow to obtain the distribution properties.

Chapter 6 presents an automated approach to produce a hierarchy of abstracts for a set of concepts inserted as an input. The goal is to provide a tool allowing for a generalization of terms without the necessity of being an expert in the domain to be generalized. Through the use of ontologies, a set of terms can be automatically generalized into the next minimally abstract level in a concept hierarchy. The algorithm for discovering these minimally abstract generalizations and the unsupervised construction of a fuzzy concept hierarchy is presented.

Chapter 7 presents a framework that goes beyond the traditional Semantic Web which has been defined mostly as a mesh or distributed databases within the World Wide Web. This chapter focuses on the development of a framework for reasoning and deduction in the Web. A Web-based approach to a decision making model for the analysis of structured database is presented. In addition, a framework to incorporate information from the Web sites into the search engine is presented as a model that will go beyond the current Semantic Web idea.

Chapter 8 illustrates a novel approach to learning user interests from the way the users interact with a document management system. The approach is based on a fuzzy conceptual representation of both documents and user interests, using information contained in an ontology. User models are constructed and updated by means of an on-line evolutionary algorithm.

Chapter 9 chapter attempts to describe some of the general issues of uncertainty in the context of the Semantic Web, in an attempt to foster research on shared epistemological frameworks for the future semantic technology. The general principles proposed accept many different technological solutions, so that a further work should deal with a process of rational inquiry on the best of them.

Chapter 10 focuses on the analysis of multimedia documents for the extraction of their semantic content. The approach is based on fuzzy algebra, as well as fuzzy ontological information. The methodologies that may lead to the creation of a semantic index are outlined. Based on the semantic index, how multimedia content may be analyzed for the extraction of semantic information in the form of thematic categorization is explained.

Northeastern University, China *Zongmin Ma*
January 2006

Contents

Part I

Probability in Ontologies and Semantic Web

BayesOWL: Uncertainty Modeling in Semantic Web Ontologies

Zhongli Ding[1,2], Yun Peng[1,3], and Rong Pan[1,4]

[1] Department of Computer Science and Electrical Engineering, University of Maryland Baltimore County, Baltimore, Maryland 21250, USA
[2] `zding1@cs.umbc.edu`
[3] `ypeng@cs.umbc.edu`
[4] `pan.rong@umbc.edu`

It is always essential but difficult to capture incomplete, partial or uncertain knowledge when using ontologies to conceptualize an application domain or to achieve semantic interoperability among heterogeneous systems. This chapter presents an on-going research on developing a framework which augments and supplements the semantic web ontology language OWL[5] for representing and reasoning with uncertainty based on Bayesian networks (BN) [26], and its application in ontology mapping. This framework, named ***BayesOWL***, has gone through several iterations since its conception in 2003 [8, 9]. *BayesOWL* provides a set of rules and procedures for direct translation of an OWL ontology into a BN directed acyclic graph (DAG), it also provides a method based on iterative proportional fitting procedure (IPFP) [19, 7, 6, 34, 2, 4] that incorporates available probability constraints when constructing the conditional probability tables (CPTs) of the BN. The translated BN, which preserves the semantics of the original ontology and is consistent with all the given probability constraints, can support ontology reasoning, both within and across ontologies as Bayesian inferences. At the present time, *BayesOWL* is restricted to translating only OWL-DL concept taxonomies into BNs, we are actively working on extending the framework to OWL ontologies with property restrictions.

If ontologies are translated to BNs, then concept mapping between ontologies can be accomplished by evidential reasoning across the translated BNs. This approach to ontology mapping is seen to be advantageous to many existing methods in handling uncertainty in the mapping. Our preliminary work on this issue is presented at the end of this chapter.

This chapter is organized as follows: Sect. 1 provides a brief introduction to semantic web[6] and discusses uncertainty in semantic web ontologies; Sect. 2

[5] `http://www.w3.org/2001/sw/WebOnt/`

[6] `http://www.w3.org/DesignIssues/Semantic.html`

Z. Ding et al.: *BayesOWL: Uncertainty Modeling in Semantic Web Ontologies*, StudFuzz **204**, 3–29 (2006)
`www.springerlink.com`

describes *BayesOWL* in detail; Sect. 3 proposes a representation in OWL of probability information concerning the entities and relations in ontologies; and Sect. 4 outlines how *BayesOWL* can be applied to automatic ontology mapping. The chapter ends with a discussion and suggestions for future research in Sect. 5.

1 Semantic Web, Ontology, and Uncertainty

People can read and understand a web page easily, but machines can not. To make web pages understandable by machines, additional semantic information needs to be attached or embedded to the existing web data. Built upon Resource Description Framework (RDF)[7], the semantic web is aimed at extending the current web so that information can be given well-defined meaning using the description logic based ontology definition language OWL, and thus enabling better cooperation between computers and people[8]. Semantic web can be viewed as a web of data that is similar to a globally accessible database.

The core of the semantic web is "ontology". In philosophy, "Ontology" is the study of the existence of entities in the universe. The term "ontology" is derived from the Greek word "onto" (means being) and "logia" (means written or spoken discourse). In the context of semantic web, this term takes a different meaning: "ontology" refers to a set of vocabulary to describe the conceptualization of a particular domain [14]. It is used to capture the concepts and their relations in a domain for the purpose of information exchange and knowledge sharing. Over the past few years, several ontology definition languages have emerged, including RDF(S), SHOE[9], OIL[10], DAML[11], DAML+OIL[12], and OWL. Among them, OWL is the newly released standard recommended by W3C[13]. A brief introduction about OWL is presented next.

1.1 OWL: Web Ontology Language

OWL, the standard web ontology language recently recommended by W3C, is intended to be used by applications to represent terms and their interrelationships. It is an extension of RDF and goes beyond its semantics. RDF is a general assertional model to represent the resources available on the web through RDF triples of "*subject*", "*predicate*" and "*object*". Each triple in RDF makes a distinct assertion, adding any other triples will not change the meaning of the existing triples. A simple datatyping model of RDF called RDF

[7] `http://www.w3.org/RDF/`
[8] `http://www.w3.org/2001/sw/`
[9] `http://www.cs.umd.edu/projects/plus/SHOE/`
[10] `http://www.ontoknowledge.org/oil/`
[11] `http://www.daml.org/`
[12] `http://www.daml.org/2001/03/daml+oil-index`
[13] `http://www.w3.org`

Schema (RDFS)[14] is used to control the set of terms, properties, domains and ranges of properties, and the "rdfs:subClassOf" and "rdfs:subPropertyOf" relationships used to define resources. However, RDFS is not expressive enough to catch all the relationships between classes and properties. OWL provides a richer set of vocabulary by further restricting on the set of triples that can be represented. OWL includes three increasingly complex variations[15]: OWL Lite, OWL DL and OWL Full.

An OWL document can include an optional ontology header and any number of class, property, axiom, and individual descriptions. In an ontology defined by OWL, a named class is described by a class identifier via "rdf:ID". An anonymous class can be described by value (owl:hasValue, owl:allValuesFrom, owl:someValuesFrom) or cardinality (owl:maxCardinality, owl:minCardinality, owl:cardinality) restriction on property (owl:Restriction); by exhaustive enumeration of all the individuals that form the instances of this class (owl:oneOf); or by logical operations on two or more other classes (owl:intersectionOf, owl:unionOf, owl:complementOf). The three logical operators correspond to AND (conjunction), OR (disjunction) and NOT (negation) in logic, they define classes of all individuals by standard set-operations of intersection, union, and complement, respectively. Three class axioms (rdfs:subClassOf, owl:equivalentClass, owl:disjointWith) can be used for defining necessary and sufficient conditions of a class.

Two kinds of properties can be defined in an OWL ontology: object property (owl:ObjectProperty) which links individuals to individuals, and datatype property (owl:DatatypeProperty) which links individuals to data values. Similar to classes, "rdfs:subPropertyOf" is used to define that one property is a subproperty of another property. There are constructors to relate two properties (owl:equivalentProperty and owl:inverseOf), to impose cardinality restrictions on properties (owl:FunctionalProperty and owl:InverseFunctionalProperty), and to specify logical characteristics of properties (owl:TransitiveProperty and owl:SymmetricProperty). There are also constructors to relate individuals (owl:sameAs, owl:sameIndividualAs, owl:differentFrom and owl:AllDifferent).

The semantics of OWL is defined based on model theory in the way analogous to the semantics of description logic (DL)[16]. With the set of vocabulary (mostly as described above), one can define an ontology as a set of (restricted) RDF triples which can be represented as an RDF graph.

1.2 Why Uncertainty?

Ontology languages in the semantic web, such as OWL and RDF(S), are based on crisp logic and thus can not handle incomplete or partial knowledge about an application domain. However, uncertainty exists in almost every aspect of

[14] `http://www.w3.org/TR/rdf-schema/`
[15] `http://www.w3.org/TR/owl-guide/`
[16] `http://www.w3.org/TR/owl-semantics/`

ontology engineering. For example, in *domain modeling*, besides knowing that "A is a subclass of B", one may also know and wishes to express that "A is a small[17] subclass of B"; or, in the case that A and B are not logically related, one may still wishes to express that "A and B are largely[18] overlapped with each other". In *ontology reasoning*, one may want to know not only if A is a subsumer of B, but also "how close of A is to B"; or, one may want to know the degree of similarity even if A and B are not subsumed by each other. Moreover, a description (of a class or an individual) one wishes to input to an ontology reasoner may be noisy and uncertain, which often leads to over-generalized conclusions in logic based reasoning. Uncertainty becomes more prevalent in *concept mapping* between two ontologies where it is often the case that a concept defined in one ontology can only find partial matches to one or more concepts in another ontology.

BayesOWL is a probabilistic framework that augments and supplements OWL for representing and reasoning with uncertainty based on Bayesian networks (BN) [26]. The basic *BayesOWL* model includes a set of structural translation rules to convert an OWL ontology into a directed acyclic graph (DAG) of a BN, and a mechanism that utilizes available probabilistic information in constructing conditional probability table (CPT) for each node in the DAG. To help understand the approach, in the remaining of this section, a brief description of BN [26] is provided.

1.3 Bayesian Networks

In the most general form, a BN of n variables consists of a directed acyclic graph (DAG) of n nodes and a number of arcs. Nodes X_i in a DAG correspond to variables, and directed arcs between two nodes represent direct causal or influential relation from one node to the other. The uncertainty of the causal relationship is represented locally by the conditional probability table (CPT) $P(X_i|\pi_i)$ associated with each node X_i, where π_i is the parent node set of X_i. Under a conditional independence assumption, the graphic structure of BN allows an unambiguous representation of interdependency between variables, which leads to one of the most important feature of BN: the joint probability distribution of $X = (X_1, \ldots, X_n)$ can be factored out as a product of the CPTs in the network (named "the chain rule of BN"):

$$P(X = x) = \prod_{i=1}^{n} P(X_i|\pi_i)$$

With the joint probability distribution, BN supports, at least in theory, any inference in the joint space. Although it has been proven that the probabilistic

[17] e.g., a probability value of 0.1 is used to quantify the degree of inclusion between A and B.

[18] e.g., a probability value of 0.9 is used to quantify the degree of overlap between A and B.

inference with general DAG structure is *NP*-hard [3], BN inference algorithms such as belief propagation [25] and junction tree [20] have been developed to explore the causal structure in BN for efficient computation.

Besides the expressive power and the rigorous and efficient probabilistic reasoning capability, the structural similarity between the DAG of a BN and the RDF graph of an OWL ontology is also one of the reasons to choose BN as the underlying inference mechanism for BayesOWL: both of them are directed graphs, and direct correspondence exists between many nodes and arcs in the two graphs.

2 The *BayesOWL* Framework

In the semantic web, an important component of an ontology defined in OWL or RDF(S) is the taxonomical concept subsumption hierarchy based on class axioms and logical relations among the concept classes. At the present time,

Table 1. Supported Constructors

Constructor	DL Syntax	Class Axiom	Logical Operator
rdfs:subClassOf	$C_1 \sqsubseteq C_2$	*	
owl:equivalentClass	$C_1 \equiv C_2$	*	
owl:disjointWith	$C_1 \sqsubseteq \neg C_2$	*	
owl:unionOf	$C_1 \sqcup ... \sqcup C_n$		*
owl:intersectionOf	$C_1 \sqcap ... \sqcap C_n$		*
owl:complementOf	$\neg C$		*

we focus our attention to OWL ontologies defined using only constructors in these two categories (as in Table 1). Constructors related to properties, individuals, and datatypes will be considered in the future.

2.1 Structural Translation

This subsection focuses on the translation of an OWL ontology file (about concept taxonomy only) into the network structure, i.e., the DAG of a BN. The task of constructing CPTs will be given in the next subsection. For simplicity, constructors for header components in the ontology, such as "owl:imports" (for convenience, assume an ontology involves only one single OWL file), "owl:versionInfo", "owl:priorVersion", "owl:backwardCompatibleWith", and "owl:incompatibleWith" are ignored since they are irrelevant to the concept definition. If the domain of discourse is treated as a non-empty collection of individuals ("owl:Thing"), then every concept class (either primitive or defined) can be thought as a countable subset (or subclass) of "owl:Thing".

Conversion of an OWL concept taxonomy into a BN DAG is done by a set of structural translation rules. The general principle underlying these rules is that all classes (specified as "subjects" and "objects" in RDF triples of the OWL file) are translated into nodes (named **concept nodes**) in BN, and an arc is drawn between two concept nodes in BN only if the corresponding two classes are related by a "predicate" in the OWL file, with the direction from the superclass to the subclass. A special kind of nodes (named **L-Nodes**) are created during the translation to facilitate modeling relations among concept nodes that are specified by OWL logical operator. These structural translation rules are summarized as follows:

(a) Every primitive or defined concept class C, is mapped into a binary variable node in the translated BN. Node C in the BN can be either "True" or "False", represented as c or $\bar{c}$, indicating whether a given instance o belongs to concept C or not.

(b) Constructor "**rdfs:subClassOf**" is modeled by a directed arc from the parent superclass node to the child subclass node. For example, a concept class C defined with superconcept classes $C_i (i = 1, ..., n)$ by "rdfs:subClassOf" is mapped into a subnet in the translated BN with one converging connection from each C_i to C, as illustrated in (Fig. 1).

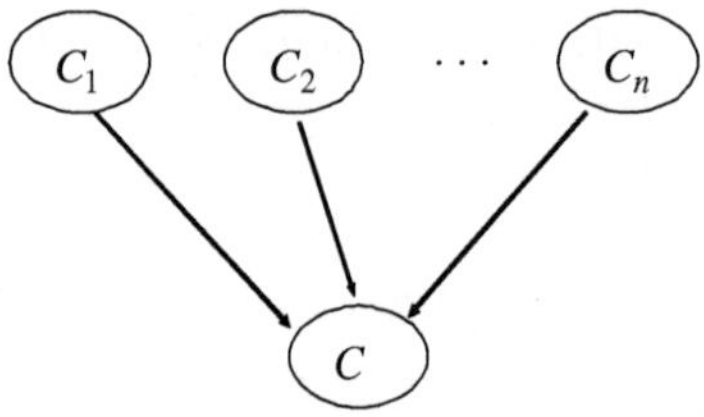

Fig. 1. "rdfs:subClassOf"

(c) A concept class C defined as the intersection of concept classes $C_i (i = 1, ..., n)$, using constructor "**owl:intersectionOf**" is mapped into a subnet (Fig. 2) in the translated BN with one converging connection from each C_i to C, and one converging connection from C and each C_i to an L-Node called "LNodeIntersection".

(d) A concept class C defined as the union of concept classes $C_i (i = 1, ..., n)$, using constructor "**owl:unionOf**" is mapped into a subnet (Fig. 3) in the translated BN with one converging connection from C to each C_i , and one converging connection from C and each C_i to an L-Node called "LNode-Union".

(e) If two concept classes C_1 and C_2 are related by constructors "**owl:complementOf**", "**owl:equivalentClass**", or "**owl:disjointWith**", then an L-Node (named "LNodeComplement", "LNodeEquivalent", "LNodeDisjoint" respectively, as in Fig. 4) is added to the translated BN, and there are directed links from C_1 and C_2 to the corresponding L-Node.

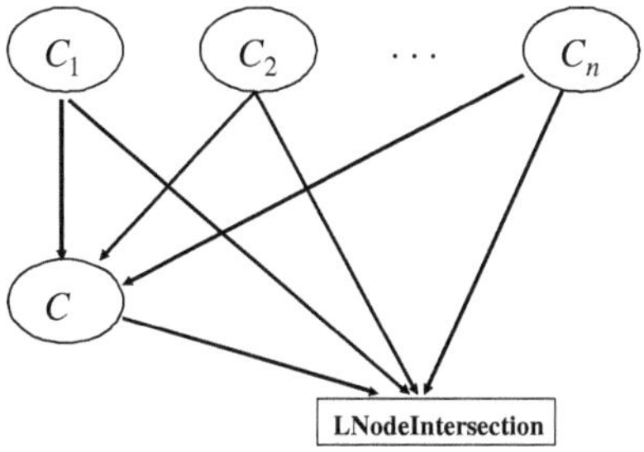

Fig. 2. "owl:intersectionOf"

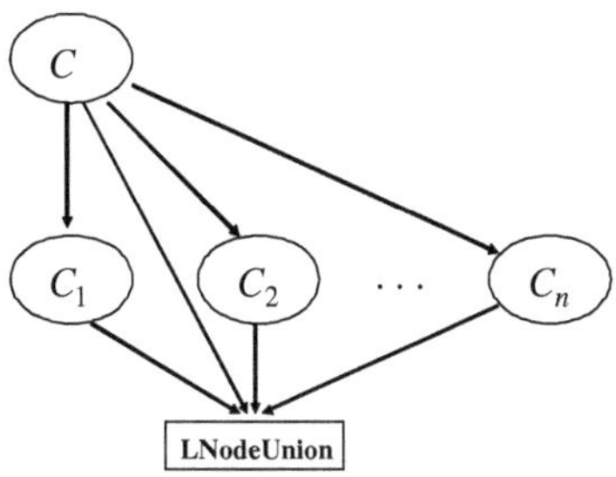

Fig. 3. "owl:unionOf"

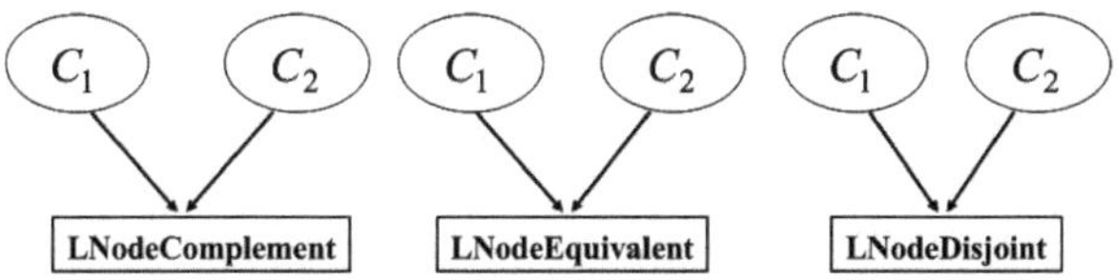

Fig. 4. "owl:complementOf, owl:equivalentClass, owl:disjointWith"

Based on rules (a) to (e), the translated BN contains two kinds of nodes: concept nodes for regular concept classes and L-Nodes which bridge concept nodes that are associated by logical relations. L-nodes are leaf nodes, with only in-arcs. With all logical relations, except "rdfs:subClassOf", handled by L-nodes, the in-arcs to a concept node can only come from its parent superclass nodes. This makes C's CPT smaller and easier to construct. L-nodes also help to avoid forming cycles in translated BN. Since L-nodes are leaves, no cycles can be formed with L-nodes. The only place where cycles can be defined for OWL taxonomies is by "rdf:subClassOf" (e.g., A is a subclass of B and B is a subclass of A). However, according to OWL semantics, all concepts involved in such a 'subclass' cycle are equivalent to each other. We can always detect this type of cycles in the pre-processing step and use rule (e), instead of rule (b), to handle the translation.

In the translated BN, all arcs are directed based on OWL statements, two concept nodes without any defined or derived relations are d-separated with each other, and two implicitly dependent concept nodes are d-connected with each other but there is no arc between them. Note that, this translation process may impose additional conditional independence to the nodes by the d-separation in the BN structure [26]. For example, consider nodes B and C, which are otherwise not related except that they both are subclasses of A. Then in the translated BN, B is conditionally independent of C, given A. Such independence can be viewed as a default relationship, which holds unless information to the contrary is provided. If dependency exists, it can be modeled by using additional nodes similar to the L-Nodes.

2.2 CPT Construction

To complete the translation the remaining issue is to assign a conditional probability table (CPT) $P(C|\pi_C)$ to each variable node C in the DAG, where π_C is the set of all parent nodes of C. As described earlier, the set of all nodes X in the translated BN can be partitioned into two disjoint subsets: concept nodes X_C which denote concept classes, and L-Nodes X_L for bridging concept nodes that are associated by logical relations.

In theory, the uncertainty information about concept nodes and their relations may be available in probability distributions of any arbitrary forms, our observation, however, is that it is most likely to be available from the domain experts or statistics in the forms of prior probabilities of concepts and pair-wise conditional probabilities of concepts, given a defined superclass. Therefore, the method developed in this chapter accommodates two types of probabilities with respect to a concept node $C \in X_C$: prior probability with the form $P(C)$, and conditional probability with the form $P(C|O_C \subseteq \pi_C)$ where $O_C \neq \emptyset$. Methods for utilizing probabilities in arbitrary forms and dimensions is reported elsewhere [28].

Before going into the details of constructing CPTs for concept nodes in X_C based on available probabilistic information (Subsect. 2.2.3), CPTs for the L-Nodes in X_L are discussed first.

2.2.1 CPTs for L-Nodes

CPT for an L-Node can be determined by the logical relation it represents so that when its state is "True", the corresponding logical relation holds among its parents. Based on the structural translation rules, there are five types of L-Nodes corresponding to the five logic operators in OWL: "LNodeComplement", "LNodeDisjoint", "LNodeEquivalent", "LNodeIntersection", and "LNodeUnion", their CPTs can be specified as follows:

(a) **LNodeComplement**: The complement relation between C_1 and C_2 can be realized by "LNodeComplement = True iff $c_1\bar{c_2} \vee \bar{c_1}c_2$", which leads to the CPT in Table 2;

Table 2. CPT of LNodeComplement

C1	C2	True	False
True	True	0.000	1.000
True	False	1.000	0.000
False	True	1.000	0.000
False	False	0.000	1.000

Table 3. CPT of LNodeDisjoint

C1	C2	True	False
True	True	0.000	1.000
True	False	1.000	0.000
False	True	1.000	0.000
False	False	1.000	0.000

Table 4. CPT of LNodeEquivalent

C1	C2	True	False
True	True	1.000	0.000
True	False	0.000	1.000
False	True	0.000	1.000
False	False	1.000	0.000

Table 5. CPT of LNodeIntersection

C1	C2	C	True	False
True	True	True	1.000	0.000
True	True	False	0.000	1.000
True	False	True	0.000	1.000
True	False	False	1.000	0.000
False	True	True	0.000	1.000
False	True	False	1.000	0.000
False	False	True	0.000	1.000
False	False	False	1.000	0.000

(b) **LNodeDisjoint**: The disjoint relation between C_1 and C_2 can be realized by "LNodeDisjoint = True iff $c_1\bar{c_2} \vee \bar{c_1}c_2 \vee \bar{c_1}\bar{c_2}$", which leads to the CPT in Table 3;

(c) **LNodeEquivalent**: The equivalence relation between C_1 and C_2 can be realized by "LNodeEquivalent = True iff $c_1c_2 \vee \bar{c_1}\bar{c_2}$", which leads to the CPT in Table 4;

(d) **LNodeIntersection**: The relation that C is the intersection of C_1 and C_2 can be realized by "LNodeIntersection = True iff $cc_1c_2 \vee \bar{c}\bar{c_1}c_2 \vee \bar{c}c_1\bar{c_2} \vee \bar{c}\bar{c_1}\bar{c_2}$", which leads to the CPT in Table 5. If C is the intersection of $n > 2$ classes, the 2^{n+1} entries in its CPT can be determined analogously.

Table 6. CPT of LNodeUnion

C1	C2	C	True	False
True	True	True	1.000	0.000
True	True	False	0.000	1.000
True	False	True	1.000	0.000
True	False	False	0.000	1.000
False	True	True	1.000	0.000
False	True	False	0.000	1.000
False	False	True	0.000	1.000
False	False	False	1.000	0.000

(e) **LNodeUnion**: The relation that C is the union of C_1 and C_2 can be realized by "LNodeUnion = True iff $cc_1c_2 \vee c\bar{c}_1c_2 \vee cc_1\bar{c}_2 \vee \bar{c}\bar{c}_1\bar{c}_2$", which leads to the CPT in Table 6. Similarly, if C is the union of $n > 2$ classes, then the 2^{n+1} entries in its CPT can be obtained analogously.

When the CPTs for L-Nodes are properly determined as above, and the states of all the L-Nodes are set to "True", the logical relations defined in the original ontology will be held in the translated BN, making the BN consistent with the OWL semantics. Denoting the situation in which all the L-Nodes in the translated BN are in "True" state as $\boldsymbol{\tau}$, the CPTs for the concept nodes in X_C should be constructed in such a way that $P(X_C|\boldsymbol{\tau})$, the joint probability distribution of all concept nodes in the subspace of $\boldsymbol{\tau}$, is consistent with all the given prior and conditional probabilistic constraints. This issue is difficult for two reasons. First, the constraints are usually not given in the form of CPT. For example, CPT for a concept node C with two parents A and B is in the form of $P(C|A,B)$ but a constraint may be given as $Q(C|A)$ or even $Q(C)$. Secondly, CPTs are given in the general space of $X = X_C \cup X_L$ but constraints are for the subspace of $\boldsymbol{\tau}$ (the dependencies changes when going from the general space to the subspace of $\boldsymbol{\tau}$). For the example constraint $Q(C|A)$, $P(C|A,B)$, the CPT for C, should be constructed in such a way that $P(C|A,\boldsymbol{\tau}) = Q(C|A)$.

To overcome these difficulties, an algorithm is developed to approximate these CPTs for X_C based on the **iterative proportional fitting procedure** (IPFP) [19, 7, 6, 34, 2, 4], a well-known mathematical procedure that modifies a given distribution to meet a set of constraints while minimizing *I-divergence* to the original distribution.

2.2.2 A Brief Introduction to IPFP

The iterative proportional fitting procedure (IPFP) was first published by Kruithof in [19] in 1937, and in [7] it was proposed as a procedure to estimate cell frequencies in contingency tables under some marginal constraints. In 1975, Csiszar [6] provided an IPFP convergence proof based on *I-divergence* geometry. Vomlel rewrote a discrete version of this proof in his PhD thesis [34] in 1999. IPFP was extended in [2, 4] as conditional iterative proportional fitting procedure (C-IPFP) to also take conditional distributions as constraints, and the convergence was established for the discrete case.

Definitions of *I-divergence* and *I-projection* are provided first before going into the details of IPFP.

Definition 1 (I-divergence)

Let $\mathbf{P}$ be a set of probability distributions over $X = \{X_1, ..., X_n\}$, and for P, $Q \in \mathbf{P}$, *I-divergence* (also known as *Kullback-Leibler divergence* or *Cross-entropy*, which is often used as a distance measure between two probability distributions) is defined as:

$$I(P\|Q) = \begin{cases} \sum\limits_{x \in X, P(x)>0} P(x) \log \frac{P(x)}{Q(x)} & if\ P \ll Q \\ +\infty & if\ P \not\ll Q \end{cases} \tag{1}$$

here $P \ll Q$ means P is dominated by Q, i.e.

$$\{x \in X | P(x) > 0\} \subseteq \{y \in X | Q(y) > 0\}$$

where x (or y) is an assignment of X, or equivalently:

$$\{y \in X | Q(y) = 0\} \subseteq \{x \in X | P(x) = 0\}$$

since a probability value is always non-negative. The dominance condition in (1) guarantees division by zero will not occur because whenever the denominator $Q(x)$ is zero, the numerator $P(x)$ will be zero. Note that *I-divergence* is zero if and only if P and Q are identical and *I-divergence* is non-symmetric.

Definition 2 (I-projection)

The *I_1-projection* of a probability distribution $Q \in \mathbf{P}$ on a set of probability distributions ε is a probability distribution $P \in \varepsilon$ such that the *I-divergence* "$I(P\|Q)$" is minimal among all probability distributions in ε. Similarly, the *I_2-projections* of Q on ε are probability distributions in ε that minimize the *I-divergence* "$I(Q\|P)$".

Note that *I_1-projection* is unique but *I_2-projection* in general is not. If ε is the set of all probability distributions that satisfies a set of given constraints, the *I_1-projection* $P \in \varepsilon$ of Q is a distribution that has the minimum distance from Q while satisfying all constraints [34].

Definition 3 (IPFP)

Let $X = \{X_1, X_2, ..., X_n\}$ be a space of n discrete random variables, given a consistent set of m marginal probability distributions $\{R(S_i)\}$ where $X \supseteq S_i \neq \emptyset$ and an initial probability distribution $Q_{(0)} \in \mathbf{P}$, iterative proportional fitting procedure (IPFP) is a procedure for determining a joint distribution $P(X) = P(X_1, X_2, ..., X_n) \ll Q_{(0)}$ satisfying all constraints in $\{R(S_i)\}$ by repeating the following computational process over k and $i = ((k-1) \bmod m) + 1$:

$$Q_{(k)}(X) = \begin{cases} 0 & if\ Q_{(k-1)}(S_i) = 0 \\ Q_{(k-1)}(X) \cdot \frac{R(S_i)}{Q_{(k-1)}(S_i)} & if\ Q_{(k-1)}(S_i) > 0 \end{cases} \tag{2}$$

This process iterates over distributions in $\{R(S_i)\}$ in cycle. It can be shown [34] that in each step k, $Q_{(k)}(X)$ is an I_1-*projection* of $Q_{(k-1)}(X)$ that satisfies the constraint $R(S_i)$, and $Q^*(X) = \lim_{k\to\infty} Q_{(k)}(X)$ is an I_1-*projection* of $Q_{(0)}$ that satisfies all constraints, i.e., $Q_k(X)$ converges to $Q^*(X) = P(X) = P(X_1, X_2, ..., X_n)$.

C-IPFP from [2, 4] is an extension of IPFP to allow constraints in the form of conditional probability distributions, i.e. $R(S_i|L_i)$ where S_i, $L_i \subseteq X$. The procedure can be written as:

$$Q_{(k)}(X) = \begin{cases} 0 & if\ Q_{(k-1)}(S_i|L_i) = 0 \\ Q_{(k-1)}(X) \cdot \frac{R(S_i|L_i)}{Q_{(k-1)}(S_i|L_i)} & if\ Q_{(k-1)}(S_i|L_i) > 0 \end{cases} \tag{3}$$

CIPF-P has similar convergence result [4] as IPFP and (2) is in fact a special case of (3) with $L_i = \emptyset$.

2.2.3 Constructing CPTs for Concept Nodes

Let $X = \{X_1, X_2, ..., X_n\}$ be the set of binary variables in the translated BN. As stated earlier, X is partitioned into two sets X_C and X_L, for concept nodes, and L-Nodes, respectively. As a BN, by chain rule [26] we have $Q(X) = \prod_{X_i \in X} Q(X_i|\pi_{X_i})$. Now, given a set of probability constraints in the forms of either

(a) prior or marginal constraint: $P(V_i)$; or

(b) conditional constraint: $P(V_i|O_{V_i})$ where $O_{V_i} \subseteq \pi_{V_i}$, $\pi_{V_i} \neq \emptyset$, $O_{V_i} \neq \emptyset$;

for $V_i \in X_C$. Also recall that all logical relations defined in the original ontology hold in the translated BN only if $\boldsymbol{\tau}$ is true (i.e., all variables in X_L are set to "True"), our objective is to construct CPTs $Q(V_i|\pi_{V_i})$ for each V_i in X_C such that $Q(X_C|\boldsymbol{\tau})$, the joint probability distribution of X_C in the subspace of $\boldsymbol{\tau}$, is consistent with all the given constraints. Moreover,we want $Q(X_C|\boldsymbol{\tau})$ to be as close as possible to the initial distribution, which may be set by human experts, by some default rules, or by previously available probabilistic information.

Note that all parents of V_i are concept nodes which are superclasses of V_i defined in the original ontology. The superclass relation can be encoded by letting every entry in $Q(V_i|\pi_{V_i})$ be zero (i.e., $Q(v_i|\pi_{V_i}) = 0$ and $Q(\bar{v}_i|\pi_{V_i}) = 1$)

if any of its parents is "False" in that entry. The only other entry in the table is the one in which all parents are "True". The probability distribution for this entry indicates the degree of inclusion of V_i in the intersection of all its parents, and it should be filled in such a way that is consistent with the given probabilistic constraints relevant to V_i. Construction of CPTs for all concept nodes thus becomes a constraint satisfaction problem in the scope of IPFP. However, it would be very expensive in each iteration of (2) or (3) to compute the joint distribution $Q_{(k)}(X)$ over all the variables and then decompose it into CPTs at the end. A new algorithm (called **Decomposed-IPFP** or **D-IPFP** for short) is developed to overcome this problem.

Let $Q_{(k)}(X_C|\boldsymbol{\tau})$ be a distribution projected from $Q_{(k)}(X_C, X_L)$ with $X_L = \boldsymbol{\tau}$ (that is, every L-Node B_j in X_L is set to b_j, the "True" state). Then by chain rule,

$$Q_{(k)}(X_C|\boldsymbol{\tau}) = \frac{Q_{(k)}(X_C, \boldsymbol{\tau})}{Q_{(k)}(\boldsymbol{\tau})}$$

$$= \frac{Q_{(k)}(V_i|\pi_{V_i}) \cdot \prod_{B_j \in X_L} Q_{(k)}(b_j|\pi_{B_j}) \cdot \prod_{V_j \in X_C, j \neq i} Q_{(k)}(V_j|\pi_{V_j})}{Q_{(k)}(\boldsymbol{\tau})} \tag{4}$$

Suppose all constraints can be decomposed into the form of $R(V_i|O_{V_i} \subseteq \pi_{V_i})$, that is, each constraint is *local* to the CPT for some $V_i \in X_C$. Apply (3) to $Q_{(k)}(X_C|\boldsymbol{\tau})$ with respect to constraint $R(V_i|O_{V_i})$ at step k,

$$Q_{(k)}(X_C|\boldsymbol{\tau}) = Q_{(k-1)}(X_C|\boldsymbol{\tau}) \cdot \frac{R(V_i|O_{V_i})}{Q_{(k-1)}(V_i|O_{V_i}, \boldsymbol{\tau})} \tag{5}$$

Then, substituting (4) to both sides of (5) and cancelling out all CPTs other than $Q(V_i|\pi_{V_i})$, we have our D-IPFP procedure as:

$$Q_{(k)}(V_i|\pi_{V_i}) = Q_{(k-1)}(V_i|\pi_{V_i}) \cdot \frac{R(V_i|O_{V_i})}{Q_{(k-1)}(V_i|O_{V_i}, \boldsymbol{\tau})} \cdot \alpha_{(k-1)}(\pi_{V_i}) \tag{6}$$

where $\alpha_{(k-1)}(\pi_{V_i}) = Q_{(k)}(\boldsymbol{\tau})/Q_{(k-1)}(\boldsymbol{\tau})$ is the normalization factor.

The process starts with $Q_{(0)} = P_{init}(X)$, the initial distribution of the translated BN where CPTs for L-Nodes are set as in Subsect. 2.2.1 and CPTs for concept nodes in X_C are set to some distributions consistent with the semantics of the subclass relation. At each iteration, only one table, $Q(V_i|\pi_{V_i})$, is modified. D-IPFP by (6) converges because (6) realizes (5), a direct application of (3), which has been shown to converge in [4].

It will be more complicated if some constraints cannot be decomposed into local constraints, e.g., $P(A|B)$, where A, B are non-empty subsets of X_C involving variables in multiple CPTs. Extending DIPFP to handle non-local constraints of more general form can be found in [28].

Some other general optimization methods such as simulated annealing (SA) and genetic algorithm (GA) can also be used to construct CPTs of the concept nodes in the translated BN. However, they are much more expensive and the quality of results is often not guaranteed. Experiments show that D-IPFP converges quickly (in seconds, most of the time in less than 30 iterative steps), despite its exponential time complexity in theoretical analysis. The space complexity of D-IPFP is trivial since each time only one node's CPT, not the entire joint probability table, is manipulated. Experiments also verify that the order in which the constraints are applied do not affect the solution, and the values of the initial distribution $Q_{(0)}(X) = P_{init}(X)$ (but avoid 0 and 1) do not affect the convergence.

2.3 Two Simple Translation Examples

First, to illustrate the using of **L-Nodes**, consider four concepts A, B, C, and D where A is equivalent to C, B is equivalent to D, and C and D are disjoint with each other. The translated BN according to our rules is depicted in Fig. 5 which realizes the given logical relations when all three L-nodes are set to "True". It also demonstrates that A and B are disjoint with each other as well.

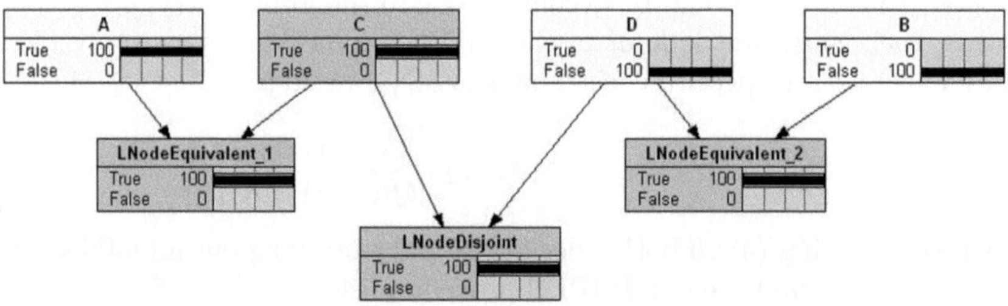

Fig. 5. Example I: Usage of L-Nodes

For the second example, a simple ontology is used here to demonstrate the validity of the approach. In this ontology, six concepts and their relations are defined as follows:

(a) *"Animal" is a primitive concept class*;
(b) *"Male", "Female", "Human" are* **subclasses** *of "Animal"*;
(c) *"Male" and "Female" are* **disjoint** *with each other*;
(d) *"Man" is the* **intersection** *of "Male" and "Human"*;
(e) *"Woman" is the* **intersection** *of "Female" and "Human"; and*
(f) *"Human" is the* **union** *of "Man" and "Woman"*.

The following probability constraints are attached to X_C = {Animal, Male, Female, Human, Man, Woman}:

(a) $P(Animal) = 0.5$
(b) $P(Male|Animal) = 0.5$
(c) $P(Female|Animal) = 0.48$
(d) $P(Human|Animal) = 0.1$
(e) $P(Man|Human) = 0.49$
(f) $P(Woman|Human) = 0.51$

When translating this ontology into BN, first the DAG of the BN is constructed (as described in Sect. 2.1), then the CPTs for L-Nodes in X_L (as described in Subsect. 2.2.1) are specified, and finally the CPTs of concept nodes in X_C are approximated by running D-IPFP. Figure 6 shows the result BN. It can be seen that, when all L-Nodes are set to "True", the conditional probability of "Male", "Female", and "Human", given "Animal", are 0.5, 0.48, and 0.1, respectively, the same as the given probability constraints. All other constraints, which are not shown in the figure due to space limitation, are also satisfied.

The CPTs of concept nodes obtained by D-IPFP are listed in Fig. 7. It can be seen that the values on the first rows in all CPTs have been changed from their initial values of (0.5, 0.5).

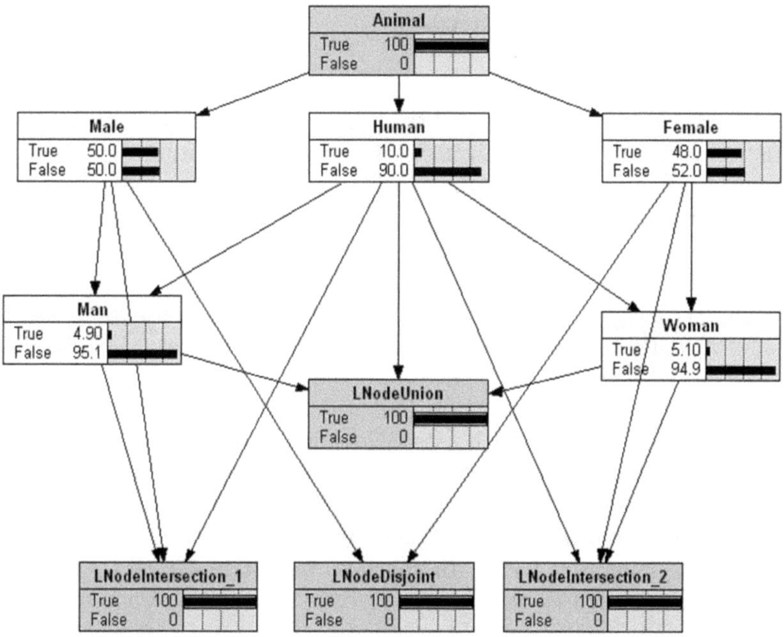

Fig. 6. Example II: DAG

Animal	
True	False
0.92752	0.07248

Animal	Human	
	True	False
True	0.18773	0.81227
False	0.0	1.0

Animal	Female	
	True	False
True	0.95469	0.04531
False	0.0	1.0

Animal	Male	
	True	False
True	0.95677	0.04323
False	0.0	1.0

Male	Human	Man	
		True	False
True	True	0.47049	0.52951
True	False	0.0	1.0
False	True	0.0	1.0
False	False	0.0	1.0

Female	Human	Woman	
		True	False
True	True	0.51433	0.48567
True	False	0.0	1.0
False	True	0.0	1.0
False	False	0.0	1.0

Fig. 7. Example II: CPT

2.4 Comparison to Related Work

Many of the suggested approaches to quantify the degree of overlap or inclusion between two concepts are based on ad hoc heuristics, others combine heuristics with different formalisms such as fuzzy logic, rough set theory, and Bayesian probability (see [32] for a brief survey). Among them, works that integrate probabilities with description logic (DL) based systems are most relevant to *BayesOWL*. This includes probabilistic extensions to $\mathcal{ALC}$ based on probabilistic logics [15, 17]; P-$\mathcal{SHOQ}$(D) [13], a probabilistic extension of $\mathcal{SHOQ}$(D) based on the notion of probabilistic lexicographic entailment; and several works on extending DL with Bayesian networks (P-CLASSIC [18] that extends CLASSIC, PTDL [36] that extends TDL (Tiny Description Logic with only "Conjunction" and "Role Quantification" operators), and the work of Holi and Hyvönen [16] which uses BN to model the degree of subsumption for ontologies encoded in RDF(S)).

The works closest to *BayesOWL* in this field are P-CLASSIC and PTDL. One difference is with CPTs. Neither of the two works has provided any mechanism to construct CPTs. In contrast, one of *BayesOWL*'s major contribution is its D-IPFP mechanism to construct CPTs from given piecewised probability constraints. Moreover, in *BayesOWL*, by using L-Nodes, the "rdfs:subclassOf" relations (or the subsumption hierarchy) are separated from other logical relations, so the in-arcs to a concept node C will only come from its parent superclass nodes, which makes C's CPT smaller and easier to construct than P-CLASSIC or PTDL, especially in a domain with rich logical relations.

Also, *BayesOWL* is not to extend or incorporate into OWL or any other ontology language or logics with probability theory, but to translate a given ontology to a BN in a systematic and practical way, and then treats ontological reasoning as probabilistic inferences in the translated BNs. Several benefits can be seen with this approach. It is non-intrusive in the sense that neither OWL nor ontologies defined in OWL need to be modified. Also, it is flexible, one can translate either the entire ontology or part of it into BN depending on the needs. Moreover, it does not require availability of complete conditional probability distributions, pieces of probability information can be incorporated into the translated BN in a consistent fashion. With these and other features, the cost of the approach is low and the burden to the user is minimal. One thing to emphasis is that *BayesOWL* can be easily extended to handle other ontology representation formalisms (syntax is not important, semantic matters), if not using OWL.

On the other side, to deal with vague and imprecise knowledge, research in extending description logics with fuzzy reasoning has gained some attention recently. Interested readers may refer to [23, 31, 1] for a rough picture about this topic.

2.5 Semantics

The semantics of the Bayesian network obtained can be outlined as follows.

(a) The translated BN will be associated with a joint probability distribution $P'(X_C)$ over the set of concept nodes X_C, and $P'(X_C) = P(X_C|\boldsymbol{\tau})$ (which can be computed by first getting the product of all the CPTs in the BN, and then marginalizing it to the subspace of $\boldsymbol{\tau}$), on top of the standard description logic semantics. A description logic interpretation $I = (\Delta^I, .^I)$ consists of a non-empty domain of objects Δ^I and an interpretation function $.^I$. This function maps every concept to a subset of Δ^I, every role and attribute to a subset of $\Delta^I \times \Delta^I$, and every individual to an object of Δ^I. An interpretation I is a model for a concept C if C^I is non-empty, and C is said "satisfiable". Besides this description logic interpretation $I = (\Delta^I, .^I)$, in *BayesOWL* semantics, there is a function P to map each object $o \in \Delta^I$ to a value between 0 and 1, $0 \leq P(o) \leq 1$, and $\sum P(o) = 1$, for all $o \in \Delta^I$. This is the probability distribution over all the domain objects. For a class C: $P(C) = \sum P(o)$ for all $o \in C$. If C and D are classes and $C \subseteq D$, then $P(C) \leq P(D)$. Then, for a node V_i in X_C, $P'(V_i) = P(V_i|\boldsymbol{\tau})$ represents the probability distribution of an arbitrary object belonging (and not belonging) to the concept represented by V_i.

(b) In the translated BN, when all the L-Nodes are set to "True", all the logical relations specified in the original OWL file will be held, which means:

1) if B is a subclass of A then "$P(b|\bar{a}) = 0 \wedge P(a|b) = 1$";

2) if B is disjoint with A then "$P(b|a) = 0 \wedge P(a|b) = 0$";

3) if A is equivalent with B then "$P(a) = P(b)$";
4) if A is complement of B then "$P(a) = 1 - P(b)$";
5) if C is the intersection of C_1 and C_2 then "$P(c|c_1, c_2) = 1 \wedge P(c|\bar{c_1}) = 0 \wedge P(c|\bar{c_2}) = 0 \wedge P(c_1|c) = 1 \wedge P(c_2|c) = 1$"; and
6) if C is the union of C_1 and C_2 then "$P(c|\bar{c_1}, \bar{c_2}) = 0 \wedge P(c|c_1) = 1 \wedge P(c|c_2) = 1 \wedge P(c_1|\bar{c}) = 0 \wedge P(c_2|\bar{c}) = 0$".
Note it would be trivial to extend 5) and 6) to general case.

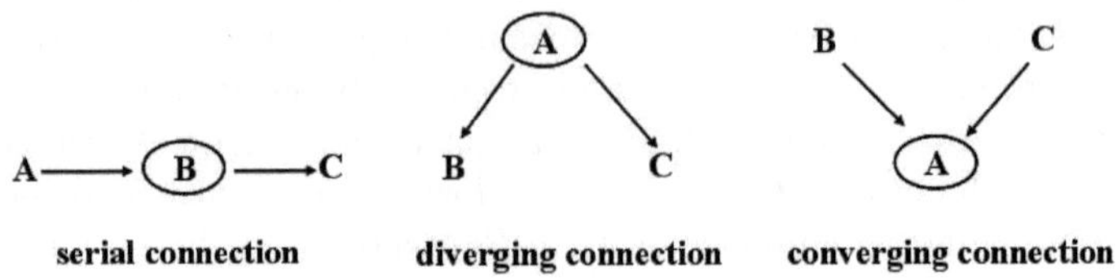

Fig. 8. Three Types of BN Connections

(c) Due to d-separation in the BN structure, additional conditional independencies may be imposed to the concept nodes in X_C in the translated BN. These are caused by the independence relations assumed in the three (serial, diverging, converging, as in Fig. 8) types of BN connections:

1) serial connection: consider A is a parent superclass of B, B is a parent superclass of C, then the probability of an object o belonging to A and belonging to C is independent if o is known to be in B;

2) diverging connection: A is the parent superclass for both B and C, then B and C is conditionally independent given A;

3) converging connection: both B and C are parent superclasses of A, then B and C are assumed to be independent if nothing about A has been known.

These independence relations can be viewed as a default relationship, which are compatible with the original ontology since there is no information to the contrary in the OWL file that defines this ontology.

2.6 Reasoning

The *BayesOWL* framework can support common ontology reasoning tasks as probabilistic reasoning in the translated BN. The follows are some of the example tasks.

(a) **Concept Satisfiability**: whether the concept represented by a description e exists. This can be answered by determining if $P(e|\boldsymbol{\tau}) = 0$.

(b) **Concept Overlapping**: the degree of the overlap or inclusion of a description e by a concept C. This can be measured by $P(e|C, \boldsymbol{\tau})$.

(c) **Concept Subsumption**: find concept C that is most similar to a given description e. This task cannot be done by simply computing the posterior $P(e|C, \boldsymbol{\tau})$, because any concept node would have higher probability than its children. Instead, a similarity measure $MSC(e, C)$ between e and C based on Jaccard Coefficient [30] is defined:

$$MSC(e, C) = P(e \cap C|\boldsymbol{\tau})/P(e \cup C|\boldsymbol{\tau}) \qquad (7)$$

This measure is intuitive and easy-to-compute. In particular, when only considering subsumers of e (i.e., $P(c|e, \boldsymbol{\tau}) = 1$), the one with the greatest MSC value is a most specific subsumer of e.

In the previous example ontology (see Fig. 6), to find the concept that is most similar to the description $e = \neg Male \sqcap Animal$, we compute the similarity measure between e and each of the nodes in $X_C = \{$Animal, Male, Female, Human, Man, Woman$\}$ using (7):

$$\begin{aligned}
&MSC(e, Animal) = 0.5004\\
&MSC(e, Male) = 0.0\\
&MSC(e, Female) = \mathbf{0.9593}\\
&MSC(e, Human) = 0.0928\\
&MSC(e, Man) = 0.0\\
&MSC(e, Woman) = 0.1019
\end{aligned}$$

This leads us to conclude that "Female" is the most similar concept to e. When a traditional DL reasoner such as Racer[19] is used, the same description would have "Animal" as the most specific subsumer, a clear over generalization.

Reasoning with uncertain input descriptions can also be supported. For example, description e' containing $P(Male) = 0.1$ and $P(Animal) = 0.7$ can be processed by inputting these probabilities as virtual evidence to the BN [27]. Class "Female" remains the most similar concept to e', but its similarity value $MSC(e', Female)$ now decreases to **0.5753**.

3 Representing Probabilities in OWL

Information about the uncertainty of the classes and relations in an ontology can often be represented as probability distributions (e.g., $P(C)$ and $P(C|D)$ mentioned earlier), which we refer to as probabilistic constraints on the ontology. These probabilities can be either provided by domain experts or learned from data.

Although not necessary, it is beneficial to represent the probabilistic constraints as OWL statements. We have developed such a representation. At the present time, we only provide encoding of two types of probabilities: priors and pair-wise conditionals. This is because they correspond naturally to classes and relations (RDF triples) in an ontology, and are most likely to be

[19] `http://www.racer-systems.com/index.phtml`

available to ontology designers. The representation can be easily extended to constraints of other more general forms if needed.

The model-theoretic semantics[20] of OWL treats the domain as a non-empty collection of individuals. If class A represents a concept, we treat it as a random binary variable of two states a and $\bar{a}$, and interpret $P(A = a)$ as the prior probability or one's belief that an arbitrary individual belongs to class A, and $P(a|b)$ as the conditional probability that an individual of class B also belongs to class A. Similarly, we can interpret $P(\bar{a})$, $P(\bar{a}|b)$, $P(a|\bar{b})$, $P(\bar{a}|\bar{b})$ and with the negation interpreted as "not belonging to".

These two types of probabilities (prior or conditional) correspond naturally to classes and relations in an ontology, and are most likely to be available to ontology designers. Currently, our translation framework can encode two types of probabilistic information into the original ontology (as mentioned earlier in Subsect. 2.2.3):

(a) prior or marginal probability $P(C)$;

(b) conditional probability $P(C|O_C)$ where $O_C \subseteq \pi_C$, $\pi_C \neq \emptyset$, $O_C \neq \emptyset$.

for a concept class C and its parent superconcept class set π_C.

We treat a probability as a kind of resource, and define two OWL classes: "PriorProb", "CondProb". A prior probability $P(C)$ of a variable C is defined as an instance of class "PriorProb", which has two mandatory properties: "hasVarible" (only one) and "hasProbValue" (only one). A conditional probability $P(C|O_C)$ of a variable C is defined as an instance of class "CondProb" with three mandatory properties: "hasCondition" (at least has one), "hasVariable" (only one), and "hasProbValue" (only one). The range of properties "hasCondition" and "hasVariable" is a defined class named "Variable", which has two mandatory properties: "hasClass" and "hasState". "hasClass" points to the concept class this probability is about and "hasState" gives the "True" (belong to) or "False" (not belong to) state of this probability.

For example, $P(c) = 0.8$, the prior probability that an arbitrary individual belongs to class C, can be expressed as follows:

```
<Variable rdf:ID="c">
   <hasClass>C</hasClass>
   <hasState>True</hasState>
</Variable>
<PriorProb rdf:ID="P(c)">
   <hasVariable>c</hasVariable>
   <hasProbValue>0.8</hasProbValue>
</PriorProb>
```

and $P(c|p1, p2, p3) = 0.8$, the conditional probability that an individual of the intersection class of $P1$, $P2$, and $P3$ also belongs to class C, can be expressed as follows:

[20] http://www.w3.org/TR/owl-semantics/direct.html

```
<Variable rdf:ID="c">
    <hasClass>C</hasClass>
    <hasState>True</hasState>
</Variable>
<Variable rdf:ID="p1">
    <hasClass>P1</hasClass>
    <hasState>True</hasState>
</Variable>
<Variable rdf:ID="p2">
    <hasClass>P2</hasClass>
    <hasState>True</hasState>
</Variable>
<Variable rdf:ID="p3">
    <hasClass>P3</hasClass>
    <hasState>True</hasState>
</Variable>
<CondProb rdf:ID="P(c|p1, p2, p3)">
    <hasCondition>p1</hasCondition>
    <hasCondition>p2</hasCondition>
    <hasCondition>p3</hasCondition>
    <hasVariable>c</hasVariable>
    <hasProbValue>0.8</hasProbValue>
</CondProb>
```

For simplicity we did not consider the namespaces in above examples. Similar to our work, [12] proposes a vocabulary for representing probabilistic relationships in an RDF graph. Three kinds of probability information can be encoded in his framework: probabilistic relations (prior), probabilistic observation (data), and probabilistic belief (posterior). And any of them can be represented using probabilistic statements which are either conditional or unconditional.

4 Concept Mapping Between Ontologies

It has become increasingly clear that being able to map concepts between different, independently developed ontologies is imperative to semantic web applications and other applications requiring semantic integration. Narrowly speaking, a mapping can be defined as a correspondence between concept A in Ontology 1 and concept B in Ontology 2 which has similar or same semantics as A. Reference [22] provides a brief survey on existing approaches for ontology-based semantic integration. Most of these works are either based on syntactic and semantic heuristics, machine learning (e.g., text classification techniques in which each concept is associates with a set of text documents that exemplify the meaning of that concept), or linguistics (spelling, lexicon relations, lexical ontologies, etc.) and natural language processing techniques.

It is often the case that, when mapping concept A defined in Ontology 1 to Ontology 2, there is no concept in Ontology 2 that is semantically identical to A. Instead, A is similar to several concepts in Ontology 2 with different degree of similarities. A solution to this so-called one-to-many problem, as suggested by [29] and [11], is to map A to the target concept B which is most similar to A by some measure. This simple approach would not work well because 1) the degree of similarity between A and B is not reflected in B and thus will not be considered in reasoning after the mapping; 2) it cannot handle the situation where A itself is uncertain; and 3) potential information loss because other similar concepts are ignored in the mapping.

To address these problems, we are pursuing an approach that combines *BayesOWL* and belief propagation between different BNs. In this approach, the two ontologies are first translated into two BNs. Concept mapping can then be processed as some form of probabilistic evidential reasoning between the two translated BNs. Our preliminary work along this direction is described in the next subsections (also refer to [10, 24] for more details and initial experimental results).

4.1 The BN Mapping Framework

In applications on large, complex domains, often separate BNs describing related subdomains or different aspects of the same domain are created, but it is difficult to combine them for problem solving – even if the interdependency relations are available. This issue has been investigated in several works, including most notably Multiply Sectioned Bayesian Network (MSBN) [35] and Agent Encapsulated Bayesian Network (AEBN) [33]. However, their results are still restricted in scalability, consistency and expressiveness. MSBN's pair-wise variable linkages are between identical variables with the same distributions, and, to ensure consistency, only one side of the linkage has a complete CPT for that variable. AEBN also requires a connection between identical variables, but allows these variables to have different distributions. Here, identical variables are the same variables reside in different BNs.

What we need in supporting mapping concepts is a framework that allows two BNs (translated from two ontologies) to exchange beliefs via variables that are similar but not identical. We illustrate our ideas by first describing how mapping shall be done for a pair of similar concepts (A from Ontology 1 to B in Ontology 2), and then discussing how such pair-wise mappings can be generalized to network to network mapping. We assume the similarity information between A and B is captured by the joint distribution $P(A, B)$.

Now we are dealing with three probability spaces: S_A and S_B for BN1 and BN2, and S_{AB} for $P(A, B)$. The mapping from A to B amounts to determine the distribution of B in S_B, given the distribution $P(A)$ in S_A under the constraint $P(A, B)$ in S_{AB}.

To propagate probabilistic influence across these spaces, we can apply Jeffrey's rule and treat the probability from the source space as soft evidence to the target space [27, 33]. This rule goes as follow. When the soft evidence on X, represented as the distribution $Q(X)$, is presented, not only $P(X)$, the original distribution of X, is changed to $Q(X)$, all other variables Y will change their distributions from $P(Y)$ to $Q(Y)$ according to (8)

$$Q(Y) = \sum_i P(Y|X_i)Q(X_i) \tag{8}$$

where the summation is over all states X_i of X.

As depicted in Fig. 9, mapping A to B is accomplished by applying Jeffrey's rule twice, first from S_A to S_{AB}, then S_{AB} to S_B. Since A in S_A is identical to A in S_{AB}, $P(A)$ in S_A becomes soft evidence $Q(A)$ to S_{AB} and by (8) the distribution of B in S_{AB} is updated to

$$Q(B) = \sum_i P(B|A_i)Q(A_i) \tag{9}$$

$Q(B)$ is then applied as soft evidence from S_{AB} to node B in S_B, updating beliefs for every other variable V in S_B by

$$\begin{aligned} Q(V) &= \sum_j P(V|B_j)Q(B_j) \\ &= \sum_j P(V|B_j) \sum_i P(B_j|A_i)Q(A_i) \\ &= \sum_j P(V|B_j) \sum_i P(B_j|A_i)P(A_i) \end{aligned} \tag{10}$$

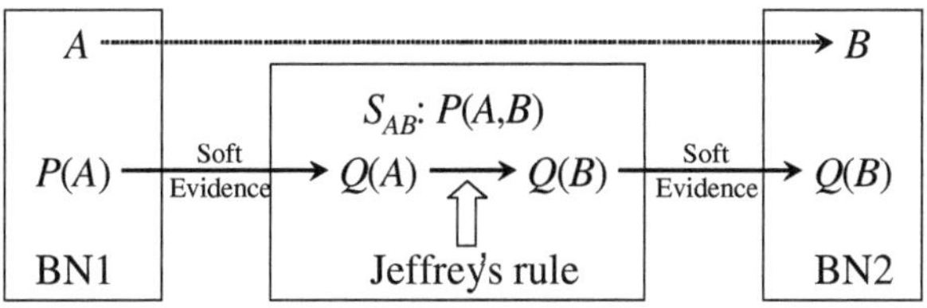

Fig. 9. Mapping Concept A to B

Back to the example in Fig. 6, where the posterior distribution of "Human", given hard evidence $\neg Male \sqcap Animal$, is $(True\ 0.102, False\ 0.898)$. Suppose we have another BN which has a variable "Adult" with marginal distribution $(True\ 0.8, False\ 0.2)$. Suppose we also know that "Adult" is similar to "Human" with conditional distribution ("T" for "True", "F" for "False")

$$P(Adult|Human) = \begin{matrix} & T & F \\ T & \multicolumn{2}{c}{\multirow{2}{*}{}} \end{matrix} \begin{array}{c} \\ T \\ F \end{array} \begin{array}{c} T \quad F \\ \begin{pmatrix} 0.7 & 0.3 \\ 0.0 & 1.0 \end{pmatrix} \end{array}$$

Mapping "Human" to "Adult" leads to a change of latter's distribution from $(True\ 0.8, False\ 0.2)$ to $(True\ 0.0714, False\ 0.9286)$ by (9). This change can then be propagated to further update believes of all other variables in the target BN by (10).

4.2 Mapping Reduction

A pair-wise linkage as described above provides a channel to propagate belief from A in BN1 to influence the belief of B in BN2. When the propagation is completed, (9) must hold between the distributions of A and B. If there are multiple such linkages, (9) must hold simultaneously for all pairs. In theory, any pair of variables between two BNs can be linked, albeit with different degree of similarities. Therefore we may potentially have $n_1 \times n_2$ linkages (n_1 and n_2 are the number of variables in BN1 and BN2, respectively). Although we can update the distribution of BN2 to satisfy all linkages by IPFP using (9) as constraints, it would be a computational formidable task.

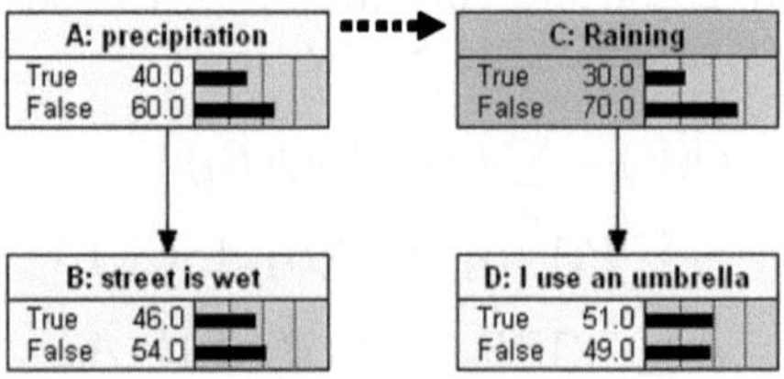

Fig. 10. Mapping Reduction Example

Fortunately, satisfying a given probabilistic relation between $P(A, B)$ does not require the utilization, or even the creation, of a linkage from A to B. Several probabilistic relations may be satisfied by one linkage. As shown in Fig. 10, we have variables A and B in BN1, C and D in BN2, and probability relations between every pair as below:

$$P(C, A) = \begin{pmatrix} 0.3 & 0.0 \\ 0.1 & 0.6 \end{pmatrix}, \qquad P(D, A) = \begin{pmatrix} 0.33 & 0.18 \\ 0.07 & 0.42 \end{pmatrix},$$

$$P(D, B) = \begin{pmatrix} 0.348 & 0.162 \\ 0.112 & 0.378 \end{pmatrix}, \qquad P(C, B) = \begin{pmatrix} 0.3 & 0.0 \\ 0.16 & 0.54 \end{pmatrix}.$$

However, we do not need to set up linkages for all these relations. As Fig. 10 depicts, when we have a linkage from A to C, all these relations are satisfied (the other three linkages are thus redundant). This is because not only beliefs on C, but also beliefs on D are properly updated by mapping A to C.

Several experiments with large BNs have shown that only a very small portions of all $n_1 \times n_2$ linkages are needed in satisfying all probability constraints. This, we suspect, is due to the fact that some of these constraints can be derived from others based on the probabilistic interdependencies among variables in the two BNs. We are currently actively working on developing a set of rules that examine the BN structures and CPTs so that redundant linkages can be identified and removed.

5 Conclusion

This chapter describes our on-going research on developing a probabilistic framework for modeling uncertainty in semantic web ontologies based on Bayesian networks. We have defined new OWL classes ("PriorProb", "CondProb", and "Variable"), which can be used to encode probability constraints for ontology classes and relations in OWL. We have also defined a set of rules for translating OWL ontology taxonomy into Bayesian network DAG and provided a new algorithm D-IPFP for efficient construction of CPTs. The translated BN is semantically consistent with the original ontology and satisfies all given probabilistic constraints. With this translation, ontology reasoning can be conducted as probabilistic inferences with potentially better, more accurate results. We are currently actively working on extending the translation to include properties, developing algorithms to support common ontology-related reasoning tasks. Encouraged by our preliminary results, we are also continuing work on ontology mapping based on *BayesOWL*. This includes formalizing concept mapping between two ontologies as probabilistic reasoning across two translated BN, and addressing the difficult issue of one-to-many mapping and its generalized form of many-to-many mapping where more than one concepts need to be mapped from one ontology to another at the same time.

The *BayesOWL*framework presented in this chapter relies heavily on the availability of probabilistic information for both ontology to BN translation and ontology mapping. This information is often not available (or only partially available) from domain experts. Learning these probabilities from data then becomes the only option for many applications. Our current focus in this direction is the approach of text classification [5, 21]. The most important and also most difficult problem in this approach is to provide high quality sample documents to each ontology class. We are exploring ontology guided search of the web for such documents.

Another interesting direction for future work is to deal with inconsistent probability information. For example, in constructing CPTs for the translated BN, the given constraints may be inconsistent with each other, also, a set of consistent constraints may itself be inconsistent with the network structure. This issue involves detection of inconsistency, identification of sources of inconsistency, and resolution of inconsistency.

References

1. Agarwal S, Hitzler P (2005) Modeling Fuzzy Rules with Description Logics. In Proceedings of Workshop on OWL Experiences and Directions. Galway, Ireland
2. Bock HH (1989) A Conditional Iterative Proportional Fitting (CIPF) Algorithm with Applications in the Statistical Analysis of Discrete Spatial Data. Bull. ISI, Contributed papers of 47th Session in Paris, 1:141–142
3. Cooper GF (1990) The Computational Complexity of Probabilistic Inference using Bayesian Belief Network. Artificial Intelligence 42:393–405
4. Cramer E (2000) Probability Measures with Given Marginals and Conditionals: I-projections and Conditional Iterative Proportional Fitting. Statistics and Decisions, 18:311–329
5. Craven M, DiPasquo D, Freitag D, McCallum A, Mitchell T, Nigam K, Slattery S (2000) Learning to Construct Knowledge Bases from the World Wide Web. Artificial Intelligence, 118(1–2): 69–114
6. Csiszar I (1975) I-divergence Geometry of Probability Distributions and Minimization Problems. The Annuals of Probability, 3(1):146–158
7. Deming WE, Stephan FF (1940) On a Least Square Adjustment of a Sampled Frequency Table when the Expected Marginal Totals are Known. Ann. Math. Statist. 11:427–444
8. Ding Z, Peng Y (2004) A Probabilistic Extension to Ontology Language OWL. In Proceedings of the 37th Hawaii International Conference on System Sciences. Big Island, HI
9. Ding Z, Peng Y, Pan R (2004) A Bayesian Approach to Uncertainty Modeling in OWL Ontology. In Proceedings of 2004 International Conference on Advances in Intelligent Systems – Theory and Applications (AISTA2004). Luxembourg-Kirchberg, Luxembourg
10. Ding Z, Peng Y, Pan R, Yu Y (2005) A Bayesian Methodology towards Automatic Ontology Mapping. In Proceedings of AAAI C&O-2005 Workshop. Pittsburgh, PA, USA
11. Doan A, Madhavan J, Domingos P, Halvey A (2003) Learning to Map between Ontologies on the Semantic Web. VLDB Journal, Special Issue on the Semantic Web
12. Fukushige Y (2004) Representing Probabilistic Knowledge in the Semantic Web. Position paper for the W3C Workshop on Semantic Web for Life Sciences. Cambridge, MA, USA
13. Giugno R, Lukasiewicz T (2002) P-*SHOQ*(D): A Probabilistic Extension of *SHOQ*(D) for Probabilistic Ontologies in the Semantic Web. INFSYS Research Report 1843–02–06, Wien, Austria
14. Gruber TR (1993) A Translation Approach to Portable Ontology Specifications. Knowledge Acquisition, 5(2):199–220
15. Heinsohn J (1994) Probabilistic Description Logics. In Proceedings of UAI-94, 311–318
16. Holi M, Hyvönen E (2004) Probabilistic Information Retrieval based on Conceptual Overlap in Semantic Web Ontologies. In Proceedings of the 11th Finnish AI Conference, Web Intelligence, Vol. 2. Finnish AI Society, Finland
17. Jaeger M (1994) Probabilistic Reasoning in Terminological Logics. In Proceedings of KR-94, 305–316
18. Koller D, Levy A, Pfeffer A (1997) P-CLASSIC: A Tractable Probabilistic Description Logic. In Proceedings of AAAI-97, 390–397

19. Kruithof R (1937) Telefoonverkeersrekening. De Ingenieur 52:E15–E25
20. Lauritzen SL, Spiegelhalter DJ (1988) Local Computation with Probabilities in Graphic Structures and Their Applications in Expert Systems. J. Royal Statistical Soc. Series B 50(2):157–224
21. McCallum A, Nigam K (1998) A Comparison of Event Models for Naive Bayes Text Classification. In AAAI-98 Workshop on "Learning for Text Categorization"
22. Noy NF (2004) Semantic Integration: A Survey Of Ontology-Based Approaches. SIGMOD Record, Special Issue on Semantic Integration, 33(4)
23. Pan JZ, Stamou G, Tzouvaras V, Horrocks I (2005) f-SWRL: A Fuzzy Extension of SWRL. In Proc. of the International Conference on Artificial Neural Networks (ICANN 2005), Special Section on "Intelligent multimedia and semantics". Warsaw, Poland
24. Pan R, Ding Z, Yu Y, Peng Y (2005) A Bayesian Network Approach to Ontology Mapping. In Proceedings of ISWC 2005. Galway, Ireland
25. Pearl J (1986) Fusion, Propagation and Structuring in Belief Networks. Artificial Intelligence 29:241–248
26. Pearl J (1988) Probabilistic Reasoning in Intelligent Systems: Networks of Plausible Inference. Morgan Kaufman, San Mateo, CA
27. Pearl J (1990) Jefferys Rule, Passage of Experience, and Neo-Bayesianism. In H.E. et al. Kyburg, Jr., editor, Knowledge Representation and Defeasible Reasoning, 245–265. Kluwer Academic Publishers
28. Peng Y, Ding Z (2005). Modifying Bayesian Networks by Probability Constraints. In the Proceedings of UAI 2005. Edinburgh, Scotland
29. Prasad S, Peng Y, Finin T (2002) A Tool For Mapping Between Two Ontologies. Poster in International Semantic Web Conference (ISWC02), Sardinia, Italy
30. van Rijsbergen CJ (1979). Information Retrieval. Lodon:Butterworths, Second Edition
31. Straccia U (2005) A Fuzzy Description Logic for the Semantic Web. In Sanchez, E., ed., Capturing Intelligence: Fuzzy Logic and the Semantic Web. Elsevier
32. Stuckenschmidt H, Visser U (2000) Semantic Translation based on Approximate Re-classification. In Proceedings of the Workshop "Semantic Approximation, Granularity and Vagueness", KR'00
33. Valtorta M, Kim Y, Vomlel J (2002) Soft Evidential Update for Probabilistic Multiagent Systems. International Journal Approximate Reasoning 29(1): 71–106
34. Vomlel J (1999) Methods of Probabilistic Knowledge Integration. PhD Thesis, Department of Cybernetics, Faculty of Electrical Engineering, Czech Technical University
35. Xiang Y (2002) Probabilistic Reasoning in Multiagent Systems: A Graphical Models Approach. Cambridge University Press
36. Yelland PM (1999) Market Analysis Using Combination of Bayesian Networks and Description Logics. Sun Microsystems Technical Report TR–99–78

Modeling Uncertainty in Semantic Web Taxonomies

Markus Holi and Eero Hyvönen

University of Helsinki, Helsinki Institute for Information Technology (HIIT),
P.O. Box 26, 00014 University of Helsinki, Finland,
`www.cs.helsinki.fi/group/seco/`
`firstname.lastname@cs.helsinki.fi`

Summary. Information retrieval systems have to deal with uncertain knowledge and query results should reflect this uncertainty in some manner. However, Semantic Web ontologies are based on crisp logic and do not provide well-defined means for expressing uncertainty. We present a new probabilistic method to approach the problem. In our method, degrees of subsumption, i.e., overlap between concepts can be modeled and computed efficiently using Bayesian networks based on RDF(S) ontologies. Degrees of overlap indicate how well an individual data item matches the query concept, which can be used as a well-defined measure of relevance in information retrieval tasks.

1 Ontologies and Information Retrieval

A key reason for using ontologies in information retrieval systems, is that they enable the representation of background knowledge about a domain in a machine understandable format. Humans use background knowledge heavily in information retrieval tasks [7]. For example, if a person is searching for documents about Europe she will use her background knowledge about European countries in the task. She will find a document about Germany relevant even if the word 'Europe' is not mentioned in it. With the help of an appropriate geographical ontology also an information retrieval system could easily make the above inference. Ontologies have in fact been used in a number of information retrieval system in recent years [14, 8, 9].

Ontologies are based on crisp logic. In the real world, however, relations between entities often include subtleties that are difficult to express in crisp ontologies. For example, most of the birds in Antarctica are penguins. Thus, if a document is annotated with the concepts 'Antarctica' and 'Bird', a human will make the inference that the document is related to Penguins. RDF(S) [2] and OWL [1] ontologies do not provide good means to make this kind of inferences.

M. Holi and E. Hyvönen: *Modeling Uncertainty in Semantic Web Taxonomies*, StudFuzz **204**, 31–46 (2006)
www.springerlink.com

Also the information system itself is a source of uncertainty. The annotation, i.e. the indexing, of the documents is often inexact or uncertain. For example, if we have a geographical ontology about the countries and areas of North-America, and we want to index photographs using it, then it is most likely that in some point we will encounter a photograph the origin of which we do not know exactly. Typically the photograph will be annotated with the concept North-America. This kind of uncertainty may also be resulted from ontology merging or evolution. When a human information searcher will encounter the annotation, she will infere that there is a high probability that the photograph is taken in the U.S.A, because it is one of the largest countries in North-America.

It would be very useful if also an information retrieval system could make the same kind of inferences and use them when constructing the result sets for queries. Notice that in the above examples the knowledge of degrees of overlap and coverage between concepts is essential for succeeding in the information retrieval task.

2 Modeling Uncertainty in Ontologies

The Venn diagram of Fig. 1 illustrates some countries and areas in the world. A crisp *partOf* meronymy cannot represent the partial overlap between the geographical area Lapland and the countries Finland, Sweden, Norway, and

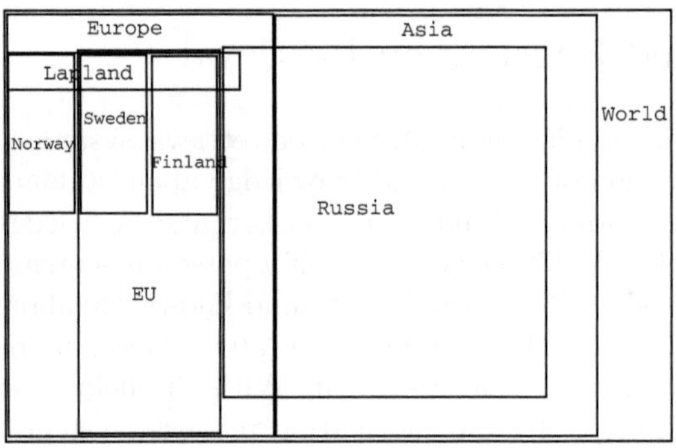

World 37*23 = 851
Europe 15*23 = 345
Asia 18*23 = 414
EU 8*21 = 168
Sweden 4*9 = 36
Finland 4*9 = 36
Norway 4*9 = 36
Lapland 13*2 = 26 Lapland&(Finland | Sweden | Norway) = 8
Lapland&EU = 16 Lapland&Russia = 2
Russia 18*19 = 342 Russia&Europe = 57 Russia&Asia = 285

Fig. 1. A Venn diagram illustrating countries, areas, their overlap, and size in the world

Russia, for example. A frequently used way to model the above situation would be to represent Lapland as the direct meronym of all the countries it overlaps, as in Fig. 2. This structure, however does not represent the situation of the map correctly, because Lapland is not subsumed by anyone of these countries. In addition, the transitivity of the subsumption relation disappears in this structure. See, for example, the relationship between Lapland and Asia. In the Venn diagram they are disjoint, but according to the taxonomy, Lapland is subsumed by Asia.

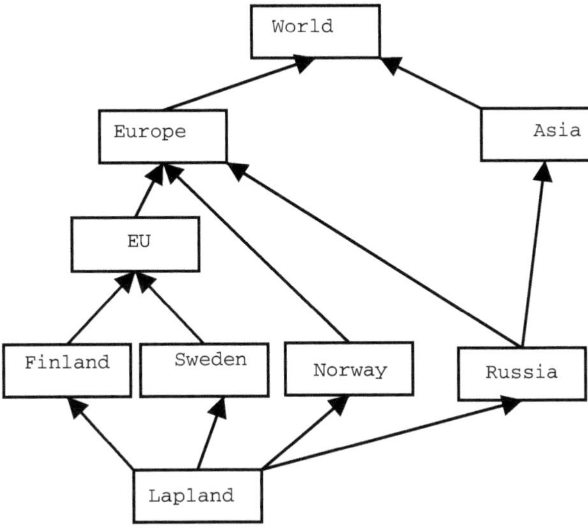

Fig. 2. A standard semantic web taxonomy based on the Venn diagram of Fig. 1

Another way would be to partition Lapland according to the countries it overlaps, as in Fig. 3. Every part is a direct meronym of both the respective country and Lapland. This structure is correct, in principle, but it too does not contain enough information to make inferences about the degrees of overlap between the areas. It does not say anything about the sizes of the different parts of Lapland, and how much they cover of the whole area of Lapland and the respective countries.

According to Fig. 1, the size of Lapland is 26 units, and the size of Finland is 36 units. The size of the overlapping area between Finland and Lapland is 8 units. Thus, 8/26 of Lapland belongs to Finland, and 8/36 of Finland belongs to Lapland. On the other hand, Lapland and Asia do not have any overlapping area, thus no part (0) of Lapland is part of Asia, and no part of Asia is part of Lapland. If we want a taxonomy to be an accurate representation of the 'map' of Fig. 1, there should be a way to make this kind of inferences based on the taxonomy.

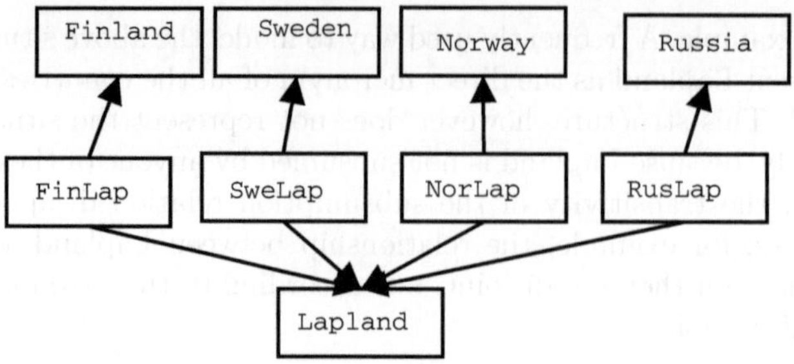

Fig. 3. Representing Lapland's overlaps by partitioning it according to the areas it overlaps. Each part is subsumed by both Lapland and the respective country

Table 1. The *overlap table* of Lapland according to Fig. 1

Selected	Referred	Overlap
Lapland	World	$26/851 = 0.0306$
	Europe	$26/345 = 0.0754$
	Asia	$0/414 = 0.0$
	EU	$16/168 = 0.0953$
	Norway	$8/36 = 0.2222$
	Sweden	$8/36 = 0.2222$
	Finland	$8/36 = 0.2222$
	Russia	$2/342 = 0.0059$

Our method enables the representation of overlap in taxonomies, and the computation of overlap between a *selected* concept and every other, i.e. *referred* concept in the taxonomy. Thus, an *overlap table* is created for the selected concept. The overlap table can be created for every concept of a taxonomy. For example, Table 1 present the overlap table of Lapland based on the the Venn diagram of Fig. 1. The Overlap column lists values expressing the mutual overlap of the selected concept and the other - referred - concepts, i.e., $Overlap = \frac{|Selected \cap Referred|}{|Referred|} \in [0, 1]$.

Intuitively, the overlap value has the following meaning: The value is 0 for disjoint concepts (e.g., Lapland and Asia) and 1, if the referred concept is subsumed by the selected one. High values lesser than one imply, that the meaning of the selected concept approaches the meaning of the referred one.

This overlap value can be used in information retrieval tasks. Assume that an ontology contains individual products manufactured in the different countries and areas of Fig. 1. The user is interested in finding objects manufactured in Lapland. The overlap values of Table 1 then tell how well the annotations "Finland", "EU", "Asia", etc., match with the query concept "Lapland" in a well-defined probabilistic sense, and the hit list can be sorted into an order of relevance accordingly.

The overlap value between the selected concept (e.g. Lapland) and the referred concept (e.g. Finland) can in fact be written as the conditional probability $P(Finland'|Lapland')$ whose interpretation is the following: If a person is interested in data records about Lapland, what is the probability that the annotation "Finland" matches her query? X' is a binary random variable such that $X' = true$ means that the annotation "X" matches the query, and $X' = false$ means that "X" is not a match. This conditional probability interpretation of overlap values will be used in section 4 of this paper. Notice that the modeling of overlap between geographical concepts, as in our example, is truly uncertain, because the exact amount of overlap is never known.

It is mathematically easy to compute the overlap tables, if a Venn diagram (the sets) is known. In practice, the Venn diagram may be difficult to create from the modeling view point, and computing with explicit sets is computationally complicated and inefficient. For these reasons our method calculates the overlap values from a taxonomic representation of the Venn diagram.

Our method consists of two parts:

1. A graphical notation by which partial subsumption and concepts can be represented in a quantified form. The notation can be represented easily in RDF(S).
2. A method for computing degrees of overlap between the concepts of a taxonomy. Overlap is quantified by transforming the taxonomy first into a Bayesian network [4].

3 Representing Overlap

In RDFS and OWL a concept, i.e. class refers to a set of individuals. Subsumption reduces essentially into the subset relationship between the sets corresponding to classes [1]. A taxonomy is therefore a set of sets and can be represented, e.g., by a Venn diagram.

If A and B are sets, then A must be in one of the following relationships to B.

1. A is a subset of B, i.e. $A \subseteq B$.
2. A partially overlaps B, i.e. $\exists x, y : (x \in A \land x \in B) \land (y \in A \land y \notin B)$.
3. A is disjoint from B, i.e. $A \cap B = \emptyset$.

Based on these relations, we have developed a simple graph notation for representing uncertainty and overlap in a taxonomy as an acyclic *overlap graph*. Here concepts are nodes, and a number called *mass* is attached to each node. The mass of concept A is a measure of the size of the set corresponding to A, i.e. $m(A) = |s(A)|$, where $s(A)$ is the set corresponding to A. A solid directed arc from concept A to B denotes crisp subsumption $s(A) \subseteq s(B)$, a dashed arrow denotes disjointness $s(A) \cap s(B) = \emptyset$, and a dotted arrow represents quantified partial subsumption between concepts, which means that the concepts partially overlap in the Venn diagram. The amount of overlap is represented by the *partial overlap value* $p = \frac{|s(A) \cap s(B)|}{|s(A)|}$.

In addition to the quantities attached to the dotted arrows, also the other arrow types have implicit overlap values. The overlap value of a solid arc is 1 (crisp subsumption) and the value of a dashed arc is 0 (disjointness). The quantities of the arcs emerging from a concept must sum up to 1. This means that either only one solid arc can emerge from a node or several dotted arcs (partial overlap). In both cases, additional dashed arcs can be used (disjointness). Intuitively, the outgoing arcs constitute a quantified partition of the concept. Thus, the dotted arrows emerging from a concept must always point to concepts that are mutually disjoint with each other.

Notice that if two concepts overlap, there must be a directed (solid or dotted) path between them. If the path includes dotted arrows, then (possible) disjointness between the concepts must be expressed explicitly using the disjointness relation. If the directed path is solid, then the concepts necessarily overlap.

For example, Fig. 4 depicts the meronymy of Fig. 1 as an overlap graph. The geographic sizes of the areas are used as masses and the partial overlap values are determined based on the Venn diagram. This graph notation is complete in the sense that any Venn diagram can be represented by it. However, sometimes the accurate representation of a Venn diagram requires the use of auxiliary concepts, which represent results of set operations over named sets, for example $s(A) \setminus s(B)$, where A and B are ordinary concepts.

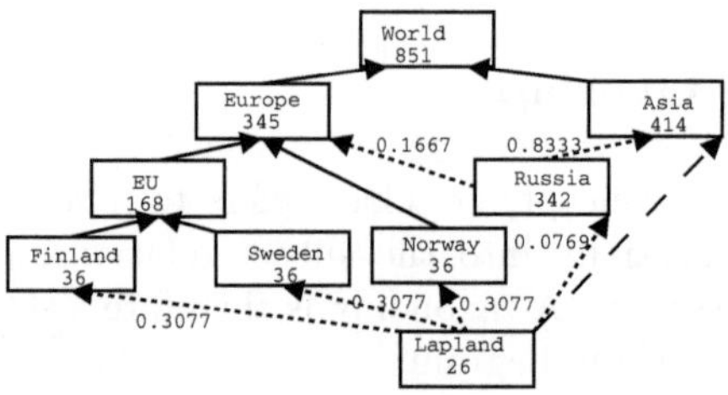

Fig. 4. The taxonomy corresponding to the Venn diagram of Fig. 1

4 Solid Path Structure

Our method creates an overlap table (cf. Fig. 1) for each concept in the taxonomy. Computing the overlaps is easiest when there are only solid arcs, i.e., complete subsumption relation, between concepts. If there is a directed solid path from A (selected) to B (referred), then overlap $o = \frac{|s(A) \cap s(B)|}{|s(B)|} = \frac{m(A)}{m(B)}$. If the solid path is directed from B to A, then $o = \frac{|s(A) \cap s(B)|}{|s(B)|} = \frac{m(B)}{m(B)} = 1$. If there is not a directed path between A and B, then $o = \frac{|s(A) \cap s(B)|}{|s(B)|} = \frac{|\emptyset|}{m(B)} = 0$.

If there is a mixed path of solid and dotted arcs between A and B, then the calculation is not as simple. Consider, for example, the relation between *Lapland* and *EU* in Fig. 4. To compute the overlap, we have to follow all the paths emerging from *Lapland*, take into account the disjoint relation between *Lapland* and *Asia*, and sum up the partial subsumption values somehow.

To exploit the simple solid arc case, a taxonomy with partial overlaps is first transformed into a *solid path structure*, in which crisp subsumption is the only relation between the concepts. The transformation is done by using to the following principle:

Transformation Principle 1 *Let A be the direct partial subconcept of B with overlap value o. In the solid path structure the partial subsumption is replaced by an additional* middle concept, *that represents $s(A) \cap s(B)$. It is marked to be the complete subconcept of both A and B, and its mass is $o \cdot m(A)$.*

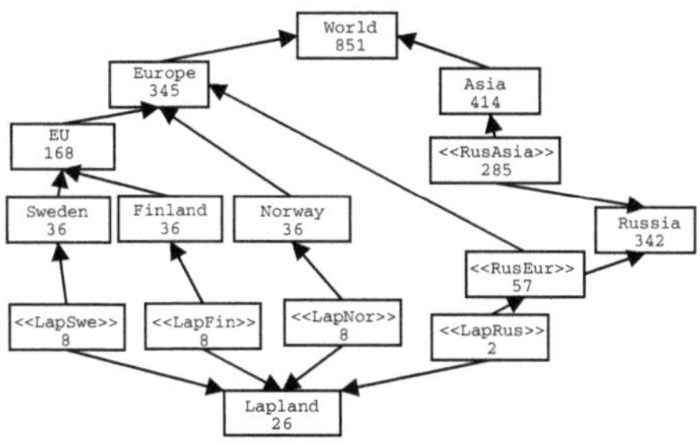

Fig. 5. The taxonomy of Fig. 4 as a solid path structure

For example, the taxonomy of Fig. 4 is transformed into the solid path structure of Fig. 5. The original partial overlaps of Lapland and Russia are transformed into crisp subsumption by using middle concepts.

The transformation is specified in algorithm 1. The algorithm processes the overlap graph T in a breadth-first manner starting from the root concept. A concept c is processed only after all of its super concepts (partial or complete) are processed. Because the graph is acyclic, all the concept will eventually be processed.

Each processed concept c is written to the solid path structure SPS. Then each arrow emerging from c is processed in the following way. If the arrow is solid, indicating subsumption, then it is written into the solid path structure as such. If the arrow is dotted, indicating partial subsumption, then a middle concept $newMc$ is added into the solid path structure. It is marked to be the complete subconcept of both c and the concept p to which the dotted arrow points in T. The mass of $newMc$ is $m(newMc) = |s(c) \cap s(p)| = o \cdot m(c)$, where o is the overlap value attached to the dotted arrow.

However, if p is connected to its superconcepts (partial or complete) with a middle concept structure, then the processing is not as simple. In that case c

```
Data: OverlapGraph T
Result: SolidPathStructure SPS
SPS := empty;
foreach concept c in T do
    foreach complete or partial direct superconcept p of c in T do
        if p connected to its superconcepts through middle concepts in SPS then
            mc := the middle concept that c overlaps;
            if c complete subconcept of p then
                mark c to be complete subconcept of mc in SPS;
            else
                newMc := middle concept representing
                s(c) ∩ s(p);
                mark newMC to be complete subconcept of c and mc in SPS;
            end
        else
            if c complete subconcept of p then
                mark c as complete subconcept of p in SPS;
            else
                newMc .= middle concept representing
                s(c) ∩ s(p);
                mark newMc to be complete subconcept of c and p in SPS;
            end
        end
    end
end
```

Algorithm 1: Creating the solid path structure

has to be connected to one of those middle concepts. The right middle concept is found by using the information conveyed in the dashed arcs emerging from c. The right middle concept mc is the one that is not subsumed by a concept that is marked to be disjoint from c in the overlap graph. This is the middle concept that c overlaps. Notice, that if the overlap graph is an accurate representation of the underlying Venn diagram, then mc is the only middle concept that fulfils the condition.

If c is a complete subconcept of p in the overlap graph T, then c is marked to be the complete subconcept of mc in SPS. If c is a partial subconcept of p in T, then it is connected to mc with a middle concept structure.

Notice, that if c was connected directly to p, instead of mc, then the information conveyed in the dashed arrows, indicating disjointness between concepts would have been lost. For example, in Fig. 5 *Lapland* was connected directly to *Russia*, then the information about the disjointness of *Lapland* and *Asia* would have been lost.

5 Computing the Overlaps

Based on the solid path structure, the overlap table values o for a selected concept A and a referred concept B could be calculated by the algorithm 2, where notation X_s denotes the set of (sub)concepts subsumed by the concept X.

The overlap table for A could be implemented by going through all the concepts of the graph and calculating the overlap value according to the above algorithm. However, because the overlap values between concepts can be interpreted as conditional probabilities, we chose to use the solid path structure as a Bayesian network topology. In the Bayesian network the boolean random variable X' replaces the concept X of the solid path structure. The efficient evidence propagation algorithms developed for Bayesian networks [4] to take care of the overlap computations. Furthermore, we saw a Bayesian representation of the taxonomy valuable as such. The Bayesian network could be used for example in user modelling [13].

if A *subsumes* B **then**
    $o := 1$
else
    $C = A_s \cap B_s$
    if $C = \emptyset$ **then**
        $o := 0$
    else
        $o := \frac{\sum_{c \in C} m(c)}{m(B)}$
    end
end

Algorithm 2: Computing the overlap

Recall from section 2 that if A is the selected concept and B is the referred one, then the overlap value o can be interpreted as the conditional probability

$$P(B' = true | A' = true) = \frac{|s(A) \cap s(B)|}{|s(B)|} = o, \tag{1}$$

where $s(A)$ and $s(B)$ are the sets corresponding to the concepts A and B. A' and B' are boolean random variables such that the value *true* means that the corresponding concept is a match to the query, i.e, the concept in question is of interest to the user. $P(B'|A')$ tells what is the probability that concept B matches the query if we know that A is a match. Notice that the Venn diagram from which $s(A)$ and $s(B)$ are taken is not interpreted as a probability space, and the elements of the sets are not interpreted as elementary outcomes of some random phenomenon. The overlap value between $s(A)$ and $s(B)$ is used merely as a means for determining the conditional probability defined above.

The joint probability distribution of the Bayesian network is defined by conditional probability tables (CPT) $P(A'|B'_1, B'_2, \ldots B'_n)$ for nodes with parents $B'_i, i = 1 \ldots n$, and by prior marginal probabilities set for nodes without parents. The CPT $P(A'|B'_1, B'_2, \ldots B'_n)$ for a node A' can be constructed by enumerating the value combinations (true/false) of the parents $B'_i, i = 1 \ldots n$, and by assigning:

$$P(A' = true|B'_1 = b_1, \ldots B'_n = b_n) = \frac{\sum\limits_{i \in \{i:b_i = true\}} m(B_i)}{m(A)} \tag{2}$$

The value for the complementary case $P(A' = false|B'_1 = b_1, \ldots B'_n = b_n)$ is obtained simply by subtracting from 1. The above formula is based on the above definition of conditional probability, and algorithm 2. The intuition behind the formula is the following. If a user is interested in Sweden and in Finland, then she is interested both in data records about Finland and in data records about Sweden. The set corresponding to this is $s(Finland) \cup s(Sweden)$. In terms of the OG this is written as $m(Finland) + m(Sweden)$. In the Bayesian network both Finland and Sweden will be set "true". Thus, the bigger the number of European countries that the user is interested in, the bigger the probability that the annotation "Europe" matches her query, i.e., $P(Europe' = true|Sweden' = true, Finland' = true) > P(Europe' = true|Finland' = true)$.

If A' has no parents, then $P(A' = true) = \lambda$, where λ is a very small non-zero probability, because we want the posterior probabilities to result from conditional probabilities only, i.e., from the overlap information.

The whole overlap table of a concept can now be determined efficiently by using the Bayesian network with its conditional and prior probabilities. By instantiating the nodes corresponding to the selected concept and the concepts subsumed by it as evidence (their values are set "true"), the propagation algorithm returns the overlap values as posterior probabilities of nodes. The query results can then be ranked according to these posterior probabilities.

Notice that when using the Bayesian network in the above way, a small inaccuracy is attached to each value as the result of the λ prior probability that was given to the parentless variables. This error approaches zero as λ approaches zero. Despite this small inaccuracy we decided to define the Bayesian network in the above manner for the following reasons.

First, to be able to easily use the the solid path structure as the topology of the Bayesian network. The CPTs can be calculated directly based on the masses of the concepts. Second, with this definition the Bayesian evidence propagation algorithm returns the overlap values readily as posterior probabilities. We experimented with various ways to construct a Bayesian network according to probabilistic interpretations of the Venn diagram. However, none of these constructions answered to our needs as well as the construction described above.

Third, in the solid path structure d-separation indicates disjointness between concepts. We see this as a useful characteristic, because it makes the simultaneous selection of two or more disjointed concepts possible.

6 Implementation

The presented method has been implemented as a proof-of-concept.

6.1 Overlap Graph

Overlap graphs are represented as RDF(S) ontologies in the following way. Concepts are represented as RDFS classes[1] The concept masses are represented using a special *Mass* class. It has two properties, subject and mass that tell the concept resource in question and mass as a numeric value, respectively. The subsumption relation can be implemented with a property of the users choice. Partial subsumption is implemented by a special *PartialSubsumption* class with three properties: subject, object and overlap. The subject property points to the direct partial subclass, the object to the direct partial superclass, and overlap is the partial overlap value. The disjointness arc is implemented by the *disjointFrom* property used in OWL.

6.2 Overlap Computations

The architecture of the implementation can be seen in Fig. 6. The input of the implementation is an RDF(S) ontology, the URI of the root node of the overlap graph, and the URI of the subsumption property used in the ontology. Additionally, also an RDF data file that contains data records annotated according to the ontology may be given. The output is the overlap tables for every concept in the taxonomy extracted from the input RDF(S) ontology. Next, each submodule in the system is discussed briefly.

The *preprocessing* module transforms the taxonomy into a predefined standard form. If an RDF data file that contains data records annotated according to the ontology is given as optional input, then the preprocessing module determines the mass of each concept in the taxonomy based on these annotations. The mass is the number of data records annotated to the concept directly or indirectly. The quantification principle is illustrated in Fig. 7.

The *transformation* module implements the transformation algorithm, and defines the CPTs of the resulting Bayesian network. In addition to the Bayesian network, it creates an RDF graph with an identical topology, where nodes are classes and the arcs are represented by the *rdf:subClassOf* property. This graph will be used by the *selection* module that expands the selection to

[1] Actually, any resources including instances could be used to represent concepts.

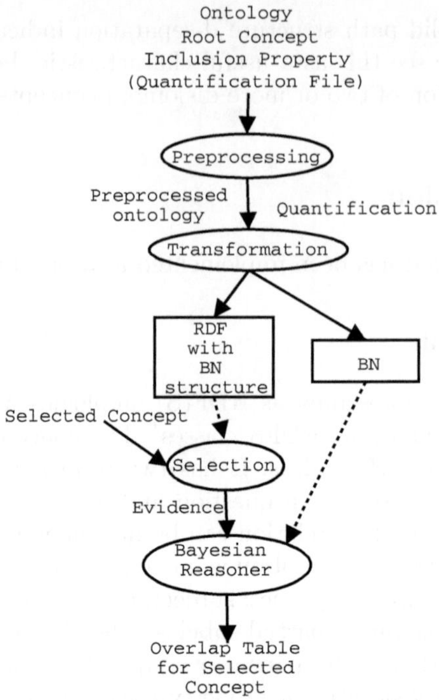

Fig. 6. The architecture of the implementation

include the concepts subsumed by the selected one, when using the Bayesian network.

The *Bayesian reasoner* does the evidence propagation based on the selection and the Bayesian network. The selection and Bayesian reasoner modules are operated in a loop, where each concept in the taxonomy is selected one after the other, and the overlap table is created.

The *preprocessing*, *transformation*, and selection modules are implemented with SWI-Prolog[2]. The Semantic Web package is used. The *Bayesian reasoner* module is implemented in Java, and it uses the Hugin Lite 6.3[3] through its Java API.

7 Discussion

7.1 Related Work

The problem of representing uncertain or vague inclusion in ontologies and taxonomies has been tackled also by using methods of fuzzy logic [21] and

[2] http://www.swi-prolog.org/

[3] http://www.hugin.com/

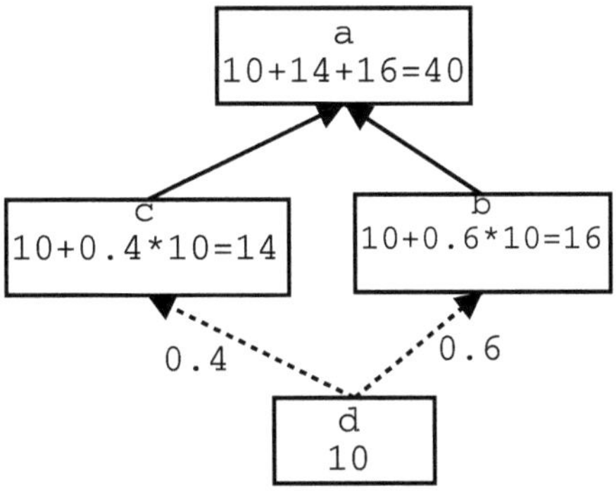

Fig. 7. Quantification of concepts. The number of direct instances of each concept is 10. In the case of partial subsumption, only a part of the mass of the subconcept is taken as the mass of the superconcept

rough sets [19, 10]. With the rough sets approach only a rough, egg-yolk representation of the concepts can be created [19]. Fuzzy logic, allows for a more realistic representation of the world.

Straccia [18] presents a fuzzy extension to the description logic SHOIN(D)corresponding to the ontology description language OWL DL. It enables the representation of fuzzy subsumption for example.

Widyantoro and Yen [20] have created a domain-specific search engine called PASS. The system includes an interactive query refinement mechanism to help to find the most appropriate query terms. The system uses a fuzzy ontology of term associations as one of the sources of its knowledge to suggest alternative query terms. The ontology is organised according to narrower-term relations. The ontology is automatically built using information obtained from the system's document collections.

The fuzzy ontology of Widyantoro and Yen is based on a set of documents, and works on that document set. However, our focus is on building taxonomies that can be used, in principle, with any data record set. The automatic creation of ontologies is an interesting issue by itself, but it is not considered in this paper. At the moment, better and richer ontologies can be built by domain specialists than by automated methods.

The fuzzy logic approach is criticised because of the arbitrariness in finding the numeric values needed and mathematical indefiniteness [19]. In addition, the representation of disjointness between concepts of a taxonomy seems to be difficult with the tools of fuzzy logic. For example, the relationships between Lapland, Russia, Europe, and Asia are very easily handled probabilistically, but in a fuzzy logic based taxonomy, this situation seems complicated. There is not a readily available fuzzy logic operation that could determine that if Lapland partly overlaps Russia, and is disjoint from Asia, then the fuzzy inclusion value between Europe and $Lapland \cap Russia$ is 1 even though Russia is only a fuzzy part of Europe.

We chose to use crisp set theory and Bayesian networks, because of the sound mathematical foundations they offer. The set theoretic approach also gives us means to overcome to a large degree the problem of arbitrariness. The calculations are simple, but still enable the representation of overlap and vague subsumption between concepts. The Bayesian network representation of a taxonomy is useful not only for the matching problem we discussed, but can also be used for other reasoning tasks [13].

Ng [17, 16] presents methods to combine probabilistic information with logic programming. This is called probabilistic logic programming. In principle we could have also created a probabilistic logic database for the taxonomy with algorithm 2. However, this would be inefficient with large ontologies, because all the possible concept combinations would have to be taken into account and encoded in the database.

Ding and Peng [3] present principles and methods to convert an OWL ontology into a Bayesian network. Their methods are based on probabilistic extensions to description logics. For more information on these extensions, see [12, 5]. The approach has some differences to ours. First, their aim is to create a method to transform any OWL ontology into a Bayesian network. Our goal is not to transform existing ontologies into Bayesian networks, but to create a method by which overlap between concepts could be represented and computed from a taxonomic structure. However, we designed the overlap graph and its RDF(S) implementation so, that it is possible, quite easily, to convert an existing crisp taxonomy to our extended notation. Second, in the approach of Ding and Peng, probabilistic information must be added to the ontology by the human modeller that needs to know probability theory. In our approach, the taxonomies can be constructed without virtually any knowledge of probability theory or Bayesian networks.

Also other approaches for combining Bayesian networks and ontologies exist. Gu [6] present a Bayesian approach for dealing with uncertain contexts. In this approach probabilistic information is represented using OWL. Probabilities and conditional probabilities are represented using classes constructed for these purposes. Mitra [15] presents a probabilistic ontology mapping tool. In this approach the nodes of the Bayesian network represents matches between pairs of classes in the two ontologies to be mapped. The arrows of the BN are dependencies between matches.

Kauppinen and Hyvönen [11] present a method for modeling partial overlap between versions of a concept that changes over long periods of time. The approach differs from ours in that we are interested in modelling degrees of overlap between different concepts in a single point of time.

7.2 Lessons Learned

Overlap graphs are simple and can be represented in RDF(S) easily. Using the notation does not require knowledge of probability or set theory. The concepts can be quantified automatically, based on data records annotated according to the ontology, for example. The notation enables the representation of any Venn diagram, but there are set structures, which lead to complicated representations.

Such a situation arises, for example, when three or more concepts mutually partially overlap each other. In these situations some auxiliary concepts have to be used. We are considering to extend the notation so that this kind of situations could be represented better. On the other hand, we do not think such situations are frequent in real-world taxonomies.

The Bayesian network structure that is created with the presented method is only one of the many possibilities. This one was chosen, because it can be used for computing the overlap tables in a most direct manner. However, it is possible that in some situations a different Bayesian network structure would be better.

7.3 Future Work

We intend to apply the overlap calculation in various realistic information retrieval situations. Also the refinement of the taxonomy language is considered to enhance its usability. The transformation of the taxonomy to alternative Bayesian network structures is an issue of future work, as well as trying the Bayesian network as a basis for personalisation.

Acknowledgements

Our research was funded mainly by the National Technology Agency Tekes.

References

1. Smith MK, Welty C, McGuinnes DL (2003) OWL web ontology language guide. http://www.w3.org/TR/2003/CR-owl-guide-20030818/
2. Brickley D, Guha RV (2004) RDF vocabulary description language 1.0: RDF Schema. http://www.w3.org/TR/rdf-schema/
3. Ding Z, Peng Y (2004) A probabilistic extension to ontology language owl. In: Proceedings of the 37th Hawaii international conference on system sciences. Big Island, Hawaii
4. Jensen FV (2001) Bayesian networks and decision graphs. Springer-Verlag, New York Berlin Heidelberg
5. Giugno R, Lukasiewicz T (2002) P-shoq(d): A probabilistic extension of shoq(d) for probabilistic ontologies in the semantic web. INFSYS research report 1843-02-06, Technische Universität Wien
6. Gu T, Pung HK, Zhang DQ (2004) A bayesian approach for dealing with uncertain contexts. In: Advances in pervasive computing. Austrian computer society, Vienna, Austria

7. Guarino N (1998) Formal ontology and information systems. In: Proceedings of FOIS'98. IOS Press, Amsterdam
8. Hyvönen E, Junnila M, Kettula S, Mäkelä E, Saarela S, Salminen M, Syreeni A, Valo A, Viljanen K (2004) Finnish museums on the semantic web. User's perspective on museumfinland. In: Proceedings of museums and the web 2004 (MW2004). Arlington, Virginia, USA
9. Hyvönen E, Valo A, Viljanen K, Holi M (2003) Publishing semantic web content as semantically linked HTML pages. In: Proceedings of XML Finland 2003. Kuopio, Finland
10. Pawlak J (1982) Rough sets. International Journal of Information and Computers 11:341–356
11. Kauppinen T, Hyvönen E (2005) Geo-spatial reasoning over ontology changes in time. In: Proceedings of IJCAI-2005 workshop on spatial and temporal reasoning. Edinburgh, Scotland
12. Koller D, Levy A, Pfeffer A (1997) P-classic: A tractable probabilistic description logic. In: Proceedings of AAAI-97
13. Kuenzer A, Schlick C, Ohmann F, Schmidt L, Luczak H (2001) An empirical study of dynamic bayesian networks for user modeling. In: Proc. of the UM'2001 workshop on machine learning for user modeling. Sonthofen, Germany
14. Mahalingam K, Huhns MN (1999) Ontology tools for semantic reconciliation in distributed heterogeneous information environments. Intelligent automation and soft computing (special issue on distributed intelligent systems)
15. Mitra P, Noy N, Jaiswal AR (2004) Omen: A probabilistic ontology mapping tool. In: Working notes of the ISWC-04 workshop on meaning coordination and negotiation. Hiroshima, Japan
16. Ng RT, Subrahmanian VS (1993) A semantical framework for supporting subjective and conditional probabilities in deductive databases. Automated reasoning journal 10(2):191–235
17. Ng RT, Tian X (1997) Semantics, consistence and query processing of empirical deductive databases. IEEE transactions on knowledge and data engineering 9(1):32–49
18. Straccia U (2005) Towards a fuzzy description logic for the semantic web. In: Proceedings of the Second european semantic web conference. Herakleion, Crete, Greece
19. Stuckenschmidt H, Visser U. (2000) Semantic translation based on approximate re-classification. In: Proceedings of the Semantic approximation, granularity and vagueness workshop. Breckenridge, Colorado, USA
20. Widyantoro DH, Yen J (2002) A fuzzy ontology-based abstract seawrch engine and its user studies. In: Proceedings of the 10th IEEE international conference on fuzzy systems. Melbourne, Australia
21. Zadeh L (1965) Fuzzy sets. Information and control 8:338-353

A Probabilistic, Logic-Based Framework for Automated Web Directory Alignment

Henrik Nottelmann[1] and Umberto Straccia[2]

[1] Institute of Informatics and Interactive Systems, University of Duisburg-Essen, Duisburg, Germany
nottelmann@uni-duisburg.de

[2] ISTI-CNR, Pisa, Italy
straccia@isti.cnr.it

Summary. We introduces oPLMap, a formal framework for automatically learning mapping rules between heterogeneous Web directories, a crucial step towards integrating ontologies and their instances in the Semantic Web. This approach is based on Horn predicate logics and probability theory, which allows for dealing with uncertain mappings (for cases where there is no exact correspondence between classes), and can be extended towards complex ontology models. Different components are combined for finding suitable mapping candidates (together with their weights), and the set of rules with maximum matching probability is selected. Our system oPLMap with different variants has been evaluated on a large test set.

1 Introduction

While the World Wide Web has been merely a collection of linked text and multimedia documents, it is currently evolving into documents with semantics, the *Semantic Web*. In this context, ontologies, which have been studied intensively for a long time, become more and more popular. Ontologies are formal definitions of concepts and their relationships. Typically, concepts are defined by classes, which are organised hierarchically by specialization (inheritance) relationships. A simple example is a Web directory, which consists of a simple class hierarchy. For example, the concept "Modern History" in Fig. 1 is a specialization (a sub-class) of the concept "History"[3].

With the emergence of ontologies and their instances in the Semantic Web, their heterogeneity constitutes a new, crucial problem. The Semantic Web is explicitly built upon the assumption that there is no commonly used ontology

[3] RDF Schema, and the OWL family of languages (OWL Full, OWL DL and OWL Lite) [21] are becoming major ontology definition languages in the Semantic Web. The latter ones are related to *Description Logics* [1], which allow for defining also properties of instances in addition to concepts

H. Nottelmann and U. Straccia: *A Probabilistic, Logic-based Framework for Automated Web Directory Alignment*, StudFuzz **204**, 47–77 (2006)
www.springerlink.com

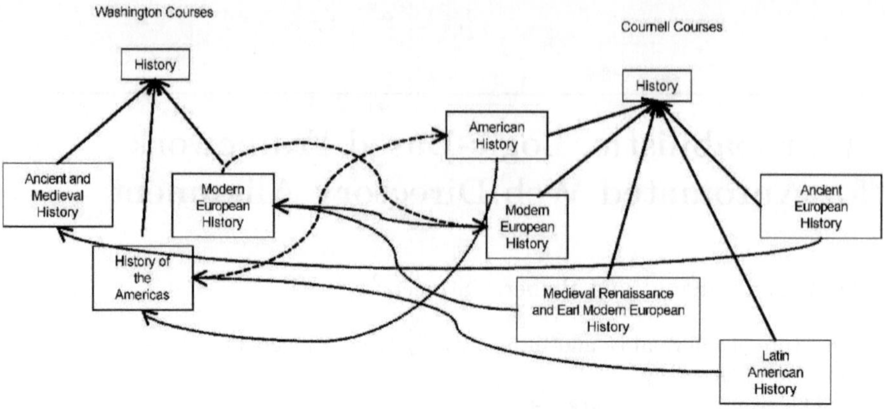

Fig. 1. The excerpt of two ontologies and class matchings

for all documents; instead, the Semantic Web will be populated by many different ontologies even for the same area. Thus, mapping/alignment between different ontologies becomes an important task. For instance, an excerpt of two "course" ontologies is given in Fig. 1. It also reports the mappings between the classes of the two ontologies. Of course, finding out these mappings automatically is desirable.

This paper proposes a new approach, called *oPLMap*, for automatically learn the mappings among tree like ontologies, e.g. web directories. Web directory alignment is the task of learning mappings between heterogeneous Web directory classes. Our approach is based on a logical framework, which is combined with probability theory (probabilistic Datalog), and aims at finding the optimum mapping (the mapping with the highest matching probability). It borrows from other approaches like GLUE [11] the idea of combining several specialized components for finding the best mapping. Using a probabilistic, logic-based framework bears some nice features: First, in many cases mappings are not absolutely correct, but hold only with a certain probability. Defining mappings by means of probabilistic rules is a natural solution to this problem. Second, classes have often attributes (called properties). These properties can easily be modelled by additional Datalog predicates. In this paper, however, we restrict to a fairly simply model where only textual content is considered.

The paper is structured as follows: The next section introduces a formal framework for learning the mappings, based on a combination of predicate logics with probability theory. Section 3 presents a theoretically founded approach for learning these mappings, where the predictions of different classifiers are combined. Our approach is evaluated on a large test bed in Sect. 4. The last section summarizes this paper, describes how this work is related to other approaches and gives an outlook over future work.

2 Web Directory Alignment

This section introduces a formal, logics-based framework for Web directory alignment. It starts from the formal framework for information exchange in [14] and extents it to a framework cable to cope with the intrinsic uncertainty of the mapping process. The framework is based on probabilistic Datalog [17], for which tools are available. The mapping process is fully automatic.

2.1 Probabilistic Datalog

In the following, we briefly describe Probabilistic Datalog (pDatalog for short) [17]. pDatalog is an extension to Datalog, a variant of predicate logic based on function-free Horn clauses. Negation is allowed, but its use is limited to achieve a correct and complete model. However, for ease of presentation we will not deal with negation in this paper. In pDatalog every fact or rule has a probabilistic weight $0 < \alpha \leq 1$ attached, prefixed to the fact or rule:

$$\alpha\ A \leftarrow B_1, \ldots, B_n\ .$$

Here, A denotes an atom (in the rule head), and $B_1, \ldots, B_n$ ($n \geq 0$) are atoms (the sub goals of the rule body). A weight $\alpha = 1$ can be omitted. In that case the rule is called *deterministic*. For ease, a fact $\alpha\ A \leftarrow$ is represented as αA. Each fact and rule can only appear once in the program, to avoid inconsistencies. The intended meaning of a rule αr is that "the probability that an instantiation of rule r is true is α". For instance, assume that we have two web directories D_1 and D_2, the class "Aeronautics_and_Astronautics" belongs to D_1, while the class "Mechanical_and_Aerospace_Engineering" belongs to D_2. Then the following rule (mapping)

```
0.1062 Mechanical_and_Aerospace_Engineering(x) ← Aeronautics_and_Astronautics(x) .
```

expresses the fact that a document about "Aeronautics_and_Astronautics" is also a document about "Mechanical_and_Aerospace_Engineering" with probability of 10.62% and, thus, establishes a bridge among the two web directories D_1 and D_2.

Formally, an *interpretation structure* is a tuple $\mathscr{I} = (\mathcal{W}, \mu)$, where $\mathcal{W}$ is a set of possible worlds and μ is a probability distribution over $\mathcal{W}$. The possible worlds are defined as follows. Given a pDatalog program P, with $H(P)$ we indicate the ground instantiation of P[4]. Then, the *deterministic part* of P is the set P_D of instantiated rules in $H(P)$ having weight $\alpha = 1$, while the *indeterministic part* of P is the set P_I of instantiated rules determined by $P_I = \{r \colon \alpha r \in H(P), \alpha < 1\}$. The set of deterministic programs of P, denoted $D(P)$ is defined as $D(P) = \{P_D \cup Y \colon Y \subseteq P_I\}$. Note that any $P' \in D(P)$ is a classical logic program. Finally, a possible world $w \in \mathcal{W}$ is the minimal model [26] of a deterministic program in $D(P)$ and is represented as the set

[4] The set of all rules that can be obtained by replacing in P the variables with constants appearing in P, i.e. the *Herbrand universe*.

of ground atoms that are true in the minimal model (also called *Herbrand model*). Now, an *interpretation* is a tuple $I = (\mathscr{I}, w)$ such that $w \in \mathcal{W}$. The truth of formulae w. r. t. an interpretation and a possible world is defined recursively as:

$$(\mathscr{I}, w) \models A \text{ iff } A \in w \ ,$$
$$(\mathscr{I}, w) \models A \leftarrow B_1, \ldots, B_n \text{ iff } (\mathscr{I}, w) \models B_1, \ldots, B_n \Rightarrow (\mathscr{I}, w) \models A \ ,$$
$$(\mathscr{I}, w) \models \alpha r \text{ iff } \mu(\{w' \in \mathcal{W} \colon (\mathscr{I}, w') \models r\}) = \alpha \ .$$

An interpretation $(\mathscr{I}, w)$ is a *model* of a pDatalog program P, denoted $(\mathscr{I}, w) \models P$, iff it entails every fact and rule in P:

$$(\mathscr{I}, w) \models P \text{ iff } (\mathscr{I}, w) \models \alpha r, \text{ for all } \alpha r \in H(P) \ .$$

In the remainder, given an n-ary atom A for predicate $\bar{A}$ and an interpretation $I = (\mathscr{I}, w)$, with A^I (an *instantiation* of A w. r. t. the interpretation A) we indicate the set of ground facts $\alpha\bar{A}(c_1, ..., c_n)$, where the ground atom $\bar{A}(c_1, ..., c_n)$ is contained in the world w, and $\mu(\{w' \in \mathcal{W} \colon (\mathscr{I}, w') \models \bar{A}(c_1, ..., c_n)\}) = \alpha$, i.e. $I \models \alpha\bar{A}(c_1, ..., c_n)$. Essentially, A^I is the set of all instantiations of A under I with relative probabilities, i.e. under I, $\bar{A}(c_1, ..., c_n)$ holds with probability α. Finally, given a ground fact αA, and a pDatalog program P, we say that P *entails* αA, denoted $P \models \alpha A$ iff in all models I of P, $I \models \alpha A$. Given a set of facts F, with say that P entails F, denoted $P \models F$, iff $P \models \alpha A$ for all $\alpha A \in F$. For ease, we will also represent an interpretation I as a set of ground facts $\{\alpha A \colon I \models \alpha A\}$. In particular, an interpretation may be seen as a pDatalog program.

2.2 Web Directories and Mappings

Web Directories

A *web directory* is a pair $\langle \mathbf{C}, \preceq \rangle$, where $\mathbf{C} = \{C_1, ..., C_n\}$ is a finite non-empty set of *classes* (or concepts, or categories) and $\preceq$ is a partial order on $\mathbf{C}$ with a top class $\top$ (for all $C \in \mathbf{C}, C \preceq \top$). The intended meaning of $C_1 \preceq C_2$ is that the class C_1 is more specific than the class C_2, i.e. all instances of C_1 are instances of C_2 (see Fig. 1).

From a logical point of view, we assume that each class C_i is an unary predicate denoting the set of object identifiers of the instances of class C_i. For ease, in the remaining of this paper, we will always assume that the object identifiers belong to a set $\mathcal{X}$. Given a web directory $(\mathbf{C}, \preceq)$ and an interpretation I, we say that I is a model of $(\mathbf{C}, \preceq)$ iff $C_1^I \models C_2^I$ whenever $C_1 \preceq C_2$. Essentially, this says that each instance of C_1 is an instance of C_2 with the same probability. Given a web directory $(\mathbf{C}, \preceq)$ and a model I of it, then the *instantiation* of $(\mathbf{C}, \preceq)$ under I, denoted $(\mathbf{C}, \preceq)^I$, is the tuple $\mathbf{C}^I = \langle C_1^I, ..., C_n^I \rangle$, i.e the tuple of all class instantiations under I[5]. Of course, an object being an instance of a class C has also attributes (sometimes called properties). Each attribute A can be modelled as a predicate $A(x, v_1, \ldots, v_l)$ indicating that the value of the attribute A of the object identified with the object identifier x is $v_1, \ldots, v_l$. For the sake of our purpose, in this paper, we use one binary relation `content` only which stores the object identifiers and the text related to them, i.e. $\texttt{content}^I \subset \mathcal{X} \times \mathcal{T}$, where $\mathcal{T}$ is the string data type. This models scenarios of Web directories where the objects are web pages. Finally, note that a web directory may easily be encoded into pDatalog as a set of rules

$$C(x) \leftarrow C'(x) \ ,$$

for all $C' \preceq C$.

Web Directory Mappings

Our goal is to automatically determine "similarity" relationships between classes of two web directories. For instance, given the web directories in Fig. 1, we would like to determine that an instance of the class "Latin American History" in the Cornell Courses Catalogue is likely an instance of the "History of the Americas" in the Washington Courses Catalogue and that "History of the Americas" is the most specific class having this property (in order to prefer the former mapping onto the "Latin American History" $\mapsto$ "History" mapping).

Theoretically, web directory mappings follow the so-called GLaV approach [25]: a *mapping* is a tuple $\mathcal{M} = (\mathbf{T}, \mathbf{S}, \Sigma)$, where $\mathbf{T}$ denotes the target (global) web directory and $\mathbf{S}$ the source (local) web directory with no relation symbol in common, and Σ is a finite set of *mapping constraints* (pDatalog rules) of the form:

$$\alpha_{j,i} \ T_j(x) \leftarrow S_i(x) \ ,$$

where T_j and S_i are target and source classes, respectively, and x is a variable ranging over object identifiers. The intended meaning of the above rules is that the class S_i of the source web directory is mapped onto the class T_j of the target web directory and the probability that this mapping is indeed true is given by $\alpha_{j,i}$. Note that a source class may be mapped onto several target classes and a target class may be the target of many source classes, i.e. we may have complex mappings

[5] One might wonder why we consider the tuple $\langle C_1^I, ..., C_n^I \rangle$ rather than the set $\{C_1^I, ..., C_n^I\}$. The reason is that in the latter case two classes may collapse together, a behaviour we want to avoid.

$$\Sigma \supseteq \{\alpha_{1,1}\ T_1(x) \leftarrow S_1(x), \alpha_{1,2}\ T_1(x) \leftarrow S_2(x), \alpha_{2,1}\ T_2(x) \leftarrow S_1(x)\} .$$

But, we do not require that we have a mapping for every target class.

For a web directory mapping $\mathcal{M} = (\mathbf{T}, \mathbf{S}, \Sigma)$ and a fixed model I for $\mathbf{S}$, a model J for $\mathbf{T}$ is a *solution* for I under $\mathcal{M}$ if and only if $\langle J, I\rangle$ (the combined interpretation over $\mathbf{T}$ and $\mathbf{S}$) is a model of Σ. The minimal solution is denoted by $J(I, \Sigma)$, the corresponding instance of $\mathbf{T}$ using interpretation $J(I, \Sigma)$ is denoted with $\mathbf{T}(I, \Sigma)$ (which is also called a minimal solution). Essentially, given a model I of $\mathbf{S}$, $\mathbf{T}(I, \Sigma)$ is the "translation/exchange" of the instances in the source web directory $\mathbf{S}^I$ into instances of the target web directory $\mathbf{T}$.

3 Learning Web Directory Mappings

Learning a web directory mapping in oPLMap consists of four steps:

1. we guess a potential web directory mapping, i.e. a set of rules Σ_k of the form $T_j(x) \leftarrow S_i(x)$ (rules without weights yet);
2. we estimate the quality of the mapping Σ_k;
3. among all possible sets Σ_k, we select the "best" web directory mapping according to our quality measure; and finally
4. the weights α for rules in the selected web directory mapping have to be estimated.

3.1 Estimating the Quality of a Mapping

Consider a target web directory $\mathbf{T} = (\{T_1, \ldots, T_t\}, \preceq_\mathbf{T})$ and a source web directory $\mathbf{S} = (\{S_1, \ldots, S_s\}, \preceq_\mathbf{S})$, and two models I of $\mathbf{S}$ and J of $\mathbf{T}$. Consider I and its minimal solution $J(I, \Sigma)$ and the corresponding instance, $\mathbf{T}(I, \Sigma)$, of $\mathbf{T}$ using interpretation $J(I, \Sigma)$. Note that $\mathbf{T}(I, \Sigma)$ contains instances of classes in $\mathbf{T}$ and each instance has its own content. For instance (see Fig. 1), consider the mapping $\mathcal{M} = (\mathbf{T}, \mathbf{S}, \Sigma)$, with $\mathbf{T}$ and $\mathbf{S}$ containing the classes

$$\mathtt{T} = \mathtt{History\_of\_the\_\ Americas}$$

$$\mathtt{S} = \mathtt{Latin\_American\_History}$$

and consider the mapping

$$\Sigma \supseteq \{\mathtt{T(X)} \leftarrow \mathtt{S(X)}\} .$$

Suppose we have a model I of the source web directory $\mathbf{S}$ with two instances identified with x1 and x2 of the class S,

$$\mathtt{S}^I = \{\mathtt{S(x1)}, \mathtt{S(x2)}\} ,$$

where their content is

$$\texttt{content(x1, ``A survey of Mexico's history...'')} \ ,$$
$$\texttt{content(x2, ``...questions of gender in Latin America...'')} \ .$$

Similarly, suppose we have a model J of the target web directory $\mathbf{T}$ with two instances identified with x3 and x4 of the class T,

$$\texttt{T}^I = \{\texttt{T(x3)}, \texttt{T(x4)}\} \ ,$$

where their content is

$$\texttt{content(x3, ``History of Latin America from colonial beginnings to the present...'')} \ ,$$
$$\texttt{content(x4, ``The American people and their culture in the modern era...'')} \ .$$

Then the minimal solution $J' = J(I, \Sigma)$ and the corresponding instance, $\mathbf{T}(I, \Sigma)$, of $\mathbf{T}$ is

$$\texttt{T}^{J'} = \{\texttt{T(x1)}, \texttt{T(x2)}\} \ .$$

Note that the facts in $\texttt{T}^J$ and $\texttt{T}^{J'}$ differ in their identifiers, but there is some "semantic overlapping" according to their content.

Our goal is to find this semantic overlapping. In particular, our goal is to find the "best" set of mapping constraints Σ, which maximises the probability $Pr(\Sigma, J, I)$ that the objects in the minimal solution $\mathbf{T}(I, \Sigma)$ under $\mathcal{M} = (\mathbf{T}, \mathbf{S}, \Sigma)$ and the objects in $\mathbf{T}^J$ are similar.

Formally, consider the minimal solution $\mathbf{T}(I, \Sigma)$ and consider a class T_j of the target web directory. With $T_j(I, \Sigma)$ we denote the restriction of $\mathbf{T}(I, \Sigma)$ to the instance of the class T_j only. Then it can be verified that Σ can be *partitioned* into sets Σ_j, where each rule in Σ_j refers to the same target class T_j (all rules in Σ_j have T_j in the head), whose minimal solutions $T_j(I, \Sigma_j)$ only contain facts for T_j:

$$\begin{aligned} \Sigma_j &= \{r : r \in \Sigma, T_j \in \mathit{head}(r)\} \ , \\ \mathbf{T}(I, \Sigma) &= \cup_{j=1}^{t} T_j(I, \Sigma_j) \ , \\ \emptyset &= T_j(I, \Sigma_j) \cap T_k(I, \Sigma_k), \ \text{if } j \neq k \ . \end{aligned}$$

Therefore, each target class can be considered independently:

$$Pr(\Sigma, J, I) = \prod_{j=1}^{t} Pr(\Sigma_j, J, I) \ .$$

We define $T_j(I, \Sigma_j)$ and T_j being similar iff $T_j(I, \Sigma_j)$ is similar to T_j and vice-versa. Thus, $Pr(\Sigma_j, J, I)$ can be computed as:

$$\begin{aligned} Pr(\Sigma_j, J, I) &= Pr(T_j | T_j(I, \Sigma_j)) \cdot Pr(T_j(I, \Sigma_j) | T_j) \\ &= Pr(T_j(I, \Sigma_j) | T_j)^2 \cdot \frac{Pr(T_j)}{Pr(T_j(I, \Sigma_j))} \\ &= Pr(T_j(I, \Sigma_j) | T_j)^2 \cdot \frac{|T_j|}{|T_j(I, \Sigma_j)|} . \end{aligned}$$

As building blocks of Σ_j, we use the sets $\Sigma_{j,i}$ containing just on rule:

$$\Sigma_{j,i} = \{\alpha_{j,i}\ T_j(x) \leftarrow S_i(x)\} . \tag{1}$$

For s source classes and a fixed j, there are also s possible sets $\Sigma_{j,i}$, and $2^s - 1$ non-empty combinations (unions) of them, forming all possible non-trivial sets Σ_j.

To simplify the notation, in the following we set $S_i = T_j(I, \Sigma_{j,i})$ for the instance derived by applying the single rule (1). For computational simplification, we assume that S_{i_1} and S_{i_2} are disjoint for $i_1 \neq i_2$. Then, for

$$\Sigma_j = \bigcup_{l=1}^{r} \Sigma_{j,i_l}$$

with indices $i_1, \ldots, i_r$, we obtain:

$$Pr(T_j(I, \Sigma_j) | T_j) = \sum_{l=1}^{r} Pr(S_{i_l} | T_j) . \tag{2}$$

Thus, to compute $Pr(\Sigma_j, J, I)$, we need to compute the $\mathcal{O}(s \cdot t)$ probabilities $Pr(S_i | T_j)$, which we will address in the next section.

3.2 Estimating the Probability of a Rule

Computing the quality of a mapping requires the probability $Pr(S_i|T_j)$, while the rule weight is $\alpha_{j,i} = Pr(T_j|S_i)$. This latter probability can easily computed from $Pr(S_i|T_j)$ as

$$Pr(T_j|S_i) = Pr(S_i|T_j) \cdot \frac{Pr(T_j)}{Pr(S_i)} = Pr(S_i|T_j) \cdot \frac{|T_j|}{|S_i|} . \tag{3}$$

Similar to GLUE [10, 11], the probability $Pr(S_i|T_j)$ is estimated by combining different classifiers $CL_1, \ldots CL_n$:

$$Pr(S_i|T_j) \approx Pr(S_i|T_j, CL_1, \ldots, CL_l) = \sum_{k=1}^{n} Pr(S_i|T_j, CL_k) \cdot Pr(CL_k) . \tag{4}$$

where the predictions $Pr(S_i|T_j, CL_k)$ is the estimate of the classifier CL_k for $Pr(S_i|T_j)$. By combining (2) and (4) we get

$$Pr(T_j(I, \Sigma_j)|T_j) = \sum_{k=1}^{n} Pr(CL_k) \cdot \sum_{l=1}^{r} Pr(S_{i_l}|T_j, CL_k) . \qquad (5)$$

The probability $Pr(CL_k)$ describes the probability that we rely on the judgment of classifier CL_k, which can for example be expressed by the confidence we have in that classifier. We simply use $Pr(CL_k) = \frac{1}{n}$ for $1 \le k \le n$, i.e. the predictions are averaged.

In practice, each classifier CL_k computes a weight $w(S_i, T_j, CL_k)$, which is the classifier's initial approximation of $Pr(S_i|T_j)$. This weight $w(S_i, T_j, CL_k)$ will be then normalized and transformed into a probability

$$Pr(S_i|T_j, CL_k) = f(w(S_i, T_j, CL_k)) ,$$

the classifier's approximation of $Pr(S_i|T_j)$. All the probabilities $Pr(S_i|T_j, CL_k)$ will then be combined together as we will see later on. The normalization process is necessary as we combine the classifier estimates, which are heterogeneous in scale. Normalization is done in two steps. First, we can consider different normalization functions:

$$\begin{aligned} f_{id}(x) &= x , \\ f_{sum}(x) &= \frac{x}{\sum_{i'} w(S_{i'}, T_j, CL_k)} , \\ f_{lin}(x) &= c_0 + c_1 \cdot x , \\ f_{log}(x) &= \frac{\exp(b_0 + b_1 \cdot x)}{1 + \exp(b_0 + b_1 \cdot x)} . \end{aligned}$$

The functions f_{id}, f_{sum} and the logistic function f_{log} return values in $[0, 1]$. For the linear function, results below zero have to mapped onto zero, and results above one have to be mapped onto one. The function f_{sum} ensures that each value is in $[0, 1]$, and that the sum equals 1. Its biggest advantage is that is does not need parameters that have to be learned. In contrast, the parameters of the linear and logistic function are learned by regression in a system-training phase. This phase is only required once, and their results can be used for learning arbitrary many web directory mappings. Of course, normalization functions can be combined. In some cases it might be useful to bring the classifier weights in the same range (using f_{sum}), and then to apply another normalization function with parameters (e.g. the logistic function).

For the final probability $Pr(S_i|T_j, CL_k)$, we have the constraint

$$0 \le Pr(S_i|T_j, CL_k) \le \frac{\min(|S_i|, |T_j|)}{|T_j|} = \min(\frac{|S_i|}{|T_j|}, 1) .$$

Thus, the normalized value (which is in $[0, 1]$) is multiplied with $\min(|S_i|/|T_j|, 1)$ in a second normalization step.

It is worth noting that some of the classifiers consider the web directories only, while others are based on the textual content, i.e. the binary relation content, which associates a text with an object. The classifiers require instances of both web directories. However, these instances do not need to describe the same objects. Below, we describe the classifiers used in this paper.

Same Class Name Stems

This binary classifier CL_S returns a weight of 1 if and only if the names of the two classes have the stem (using e.g. a Porter stemmer), and 0 otherwise:

$$w(S_i, T_j, CL_S) = \begin{cases} 1 \text{ if } S_i, T_j \text{ have same stem} , \\ 0 \text{ otherwise} . \end{cases}$$

Coordination-level Match on Class Names

This classifier CL_{N-clm} employs information retrieval (IR) techniques by applying the coordination-level match similarity function onto the class name. For this, the class names S_i and T_j are considered as bags (multi-sets) of words; the words are obtained by converting the name into lower case, splitting it into tokens, removing stop words (frequent words without semantics), and apply stemming on the remaining tokens (which maps different derivations onto a common word "stem", e.g. "computer" and "computing" onto "comput"). The prediction is computed as overlap of the resulting sets of both class names:

$$w(S_i, T_j, CL_{N-clm}) = \frac{|S_i \cap T_j|}{|S_i \cup T_j|} .$$

Coordination-level Match on Class Path Names

This classifier CL_{PN-clm} is equivalent to CL_{N-clm}, but is applied on the complete "path" of a class C. With $C \preceq C_1 \preceq \cdots \preceq C_n \preceq \top$, this is the concatenation of the names of C as well as the names of $C_1, \ldots, C_n$. This concatenation is considered as a bag of words, and the same weights as for CL_{N-clm} are computed.

kNN Classifier

A popular classifier for text and facts is kNN [35], which also employs IR techniques. For CL_{kNN}, each class S_i acts as a category, and training sets are formed from the instances of S_i:

$$Train = \bigcup_{i=1}^{s} \{(S_i, x, v) : (x, v) \in \texttt{content},\ x \in S_i\} .$$

For every instance $x \in T_j$ and its content v (i.e., the value v with $(x, v) \in$ content)[6], the k-nearest neighbours TOP_k have to be found by ranking the

[6] By abuse of notation, with $x \in T_j$ we denote that object x is an instance of class T_j, i.e. $T_j(x) \in T_j^J$. Thus, $(x, v) \in$ content is used as a shorthand for $\texttt{content}(x, v) \in \texttt{content}^J$.

values $(S_i, x', v') \in \mathit{Train}$ according to their similarity $RSV(v, v')$. The prediction weights are then computed by summing up the similarity values for all x' which are built from S_i, and by averaging these weights $\tilde{w}(x, v, S_i)$ over all instances $x \in T_j$:

$$w(S_i, T_j, CL_{kNN}) = \frac{1}{|T_j|} \cdot \sum_{(x,v)\in\texttt{content},\ x\in T_j} \tilde{w}(x, v, S_i) \ ,$$

$$\tilde{w}(x, v, S_i) = \sum_{(S_l, x'v')\in TOP_k, S_i = S_l} RSV(v, v') \ ,$$

$$RSV(v, v') = \sum_{w\in v\cap v'} Pr(w|v) \cdot Pr(w|v') \ ,$$

$$Pr(w|v) = \frac{\mathit{tf}(w, v)}{\sum_{w'\in v} \mathit{tf}(w', v)} \ ,$$

$$Pr(w|v') = \frac{\mathit{tf}(w, v')}{\sum_{w'\in v'} \mathit{tf}(w', v')} \ .$$

Here, $\mathit{tf}(w, v)$ denotes the number of times the word w appears in the string v (seen as a bag of words). The quantity $\mathit{tf}(w, v')$ is similar.

Naive Bayes Text Classifier

The classifier CL_B uses a naive Bayes text classifier [35] for text content. As for the other classifiers, each class acts as a category, and class values are considered as bags of words (with normalized word frequencies as probability estimations). For each $(x, v) \in \texttt{content}$ with $x \in T_j$, the probability $Pr(S_i|v)$ that the value v should be mapped onto S_i is computed. In a second step, these probabilities are combined by:

$$w(S_i, T_j, CL_B) = \sum_{(x,v)\in\texttt{content},\ x\in T_j} Pr(S_i|v) \cdot Pr(v) \ .$$

Again, we consider the values as bags of words. With $Pr(S_i)$ we denote the probability that a randomly chosen value in $\bigcup_k S_k$ is a value in S_i, and $Pr(w|S_i) = Pr(w|v(S_i))$ is defined as for kNN, where $v(S_i) = \bigcup_{(x,v)\in\texttt{content},\ x\in S_i} v$ is the combination of all words in all values for all objects in S_i (again considered as bags). If we assume independence of the words in a value, then we obtain:

$$Pr(S_i|v) = Pr(v|S_i) \cdot \frac{Pr(S_i)}{Pr(v)} = \frac{Pr(S_i)}{Pr(v)} \cdot \prod_{w\in v} Pr(w|S_i) \ .$$

Together, the final formula is:

$$w(S_i, T_j, CL_B) = Pr(S_i) \cdot \sum_{(x,v)\in\texttt{content},\ x\in T_j} \prod_{w\in v} Pr(w|S_i) \ .$$

If a word does not appear in the content for any object in S_i, i.e. $Pr(w|S_i) = 0$, we assume a small value to avoid a product of zero.

3.3 Exploiting the Hierarchical Structure

So far, the oPLMap learning approach does not exploit the hierarchical nature of the web directories, i.e. the partial orders $\preceq_{\mathbf{T}}$ and $\preceq_{\mathbf{S}}$. To do so, we apply additional classifiers after we have computed the prediction $w'(S_i, T_j) = Pr(S_i|T_j, CL_1, \ldots, CL_l)$ from the so far considered classifiers $CL_1, \ldots, CL_l$. The rationale of this separation is that we want to avoid cyclic dependencies between hierarchical and non-hierarchical classifiers. The predictions of the hierarchical classifiers can then be combined with the predictions of the previous classifiers as before. Formally, we introduce the set B with the best matchings, i.e. where S_i is the best attribute on which T_j can be mapped onto: $B = \{(S_i, T_j)\colon w'(S_i, T_j) \geq \max_{S'} w'(S', T_j)\}$.

Matching Parents

The binary classifier CL_P returns 1 if and only if two parents of the two source and target classes have highest matching prediction for all other $S_{i'}$ classes:

$$p_{\mathbf{S}}(C) = \{C'\colon C \preceq_{\mathbf{S}} C'\}$$
$$p_{\mathbf{T}}(C) = \{C'\colon C \preceq_{\mathbf{T}} C'\}$$

$$w(S_i, T_j, CL_P) = \begin{cases} 1 \text{ if } p_{\mathbf{S}}(S_i) \times p_{\mathbf{T}}(T_j) \cap B \neq \emptyset \,, \\ 0 \text{ otherwise .} \end{cases}$$

Matching Children

The classifier CL_C returns the amount of matching children:

$$c_{\mathbf{S}}(C) = \{C' : C' \preceq_{\mathbf{S}} C\}$$
$$c_{\mathbf{T}}(C) = \{C' : C' \preceq_{\mathbf{T}} C\}$$
$$C(S_i, T_j) = c_{\mathbf{S}}(S_i) \times c_{\mathbf{T}}(T_j)$$
$$w(S_i, T_j, CL_C) = \frac{|C(S_i,T_j) \cap B|}{|C(S_i,T_j)|} \,.$$

3.4 Additional Constraints

Additional constraints can be applied on the learned rules for improving precision. These constraints are used after the sets of rules are learned for all target classes: we remove learned rules that violate one of these constraints. These constraints are stated against the hierarchical structure of the web directories:

1. We can assume that parent-child relationships in **S** and **T** are not reversed. In other words, we assume that for $S_{i_1} \preceq_{\mathbf{S}} S_{i_2}$ and $T_{j_1} \preceq_{\mathbf{T}} T_{j_2}$, it is not possible to map S_{i_1} onto T_{j_2} and S_{i_2} onto T_{j_1} together.
2. We can assume that if a source class S_{i_2} is parent of another source class S_{i_1}, then target classes onto which S_{i_2} is mapped are parents of target classes onto which S_{i_1} are mapped. Thus, $S_{i_1} \preceq_{\mathbf{S}} S_{i_2}$ and two rules $T_{j_1}(x) \leftarrow S_{i_1}(x)$ and $T_{j_2}(x) \leftarrow S_{i_2}(x)$ implies $T_{j_1} \preceq_{\mathbf{T}} T_{j_2}$.
3. Another assumption is that there is at most one rule for the target class. This will reduce the number of rules produced, and hopefully increase the percentage of correct rules.
4. We can drop all rules whose weight $\alpha_{j,i}$ is lower than a threshold ε, e.g. with $\varepsilon = 0.1$.
5. We can rank the rules according to their weights (in decreasing order), and use the n top-ranked rules (e.g. $n = 50$).

If a constraint is violated, the rule with the lower weight will be removed.

4 Experiments

This section describes the results from the oPLMap evaluation.

4.1 Evaluation Setup

This section describes the test set (source and target instances) and the classifiers used for the experiments. It also introduces different effectiveness measurements for evaluating the learned web directory mappings (error, precision, recall). Experiments were performed on the "course catalog" test bed[7]. The Cornell University course catalog consists of 176 classes, among them 149 leaf concepts (in a maximum depth of 4), and 4,360 instances. The University of Washington contains 147 classes (141 leaf classes, maximum depth is again 4), and 6,957 instances.

Each collection is split randomly into four sub-collections of approximately the same size. The first sub-collection is always used for learning the parameters of the normalization functions (same documents in both web directories). The second sub-collection is used as source instance for learning the rules, and the third sub-collection is used as the target instance. Finally, the fourth sub-collection is employed for evaluating the learned rules (for both instances, i.e. we evaluate on parallel corpora). Rules are learned for both directions.

Each of web directory- and text-based classifiers introduced in Sect. 3.2 are used alone, plus the combinations of all of these classifiers. For the hierarchical classifiers, none, both classifiers are used alone and in combination. In every experiment, every classifier used the same normalization function from Sect. 3.2 and combinations of them.

[7] `http://anhai.cs.uiuc.edu/archive/domains/course_catalog.html`

$Pr(T_j(x) \in T_j^J)$ denotes the probability of a tuple x to be instance of the target attribute T_j in the model J of $\mathbf{T}$, i.e.: $Pr(T_j(x) \in T_j^J) = \alpha$ iff $T_j^J \models \alpha T_j(x)$ iff $\alpha T_j(x) \in T_j^J$. Often the target instance only contains deterministic data, then we have $Pr(T_j(x) \in T_j^J) \in \{0,1\}$. Similarly, $Pr(T_j(x) \in T_j(I, \Sigma_j)) \in [0,1]$ denotes the probability of a tuple x to be instance of the target attribute T_j in the minimal model $T_j(I, \Sigma_j)$ of the mapping $\mathcal{M} = (\mathbf{T}, \mathbf{S}, \Sigma)$ w.r.t. the model I of the source schema $\mathbf{S}$. Remind that $T_j(I, \Sigma_j)$ is obtained by applying all rules in Σ to the elements in the source instance $\mathbf{S}^I$ and then project the result on the target attribute T_j. Finally, the error of the mapping is defined by:

$$E(\mathcal{M}) = \frac{1}{\sum_j |U_j|} \sum_j \sum_{x \in U_j} (Pr(T_j(x) \in T_j^J) - Pr(T_j(x) \in T_j(I, \Sigma_j)))^2 ,$$

where $U_j = \{T_j(x) \in T_j^J\} \cup \{T_j(x) \in T_j(I, \Sigma_j)\}$. Furthermore, we evaluated if the learning approach computes the correct rules (neglecting the corresponding rule weights). Similar to the area of Information Retrieval [2], precision defines how many learned rules are correct, and how many correct rules are learned. In the following, R_L denotes the set of rules (without weights) returned by the learning algorithm, and R_A the set of rules (again without weights), which are the actual ones.

As we deal with hierarchical web directories, this hierarchy should also be included in the measures. Thus, for $S_1 \preceq_{\mathbf{S}} S_2$ and the rules

$$R_L = \{T_1(x) \leftarrow S_1(x)\}$$
$$R_A = \{T_1(x) \leftarrow S_2(x)\}$$

traditional precision and recall would be zero with these definitions:

$$precision_{trad} = \frac{|R_L \cap R_A|}{|R_L|} , \quad recall_{trad} = \frac{|R_L \cap R_A|}{|R_A|} .$$

However, the learned rule is too specific, but not completely wrong. The following definition takes that into consideration. With $d(C_1, C_2)$, we denote the distance between two classes C_1 and C_2 (from the same web directory) in the hierarchy:

$$d(C_1, C_2) = \begin{cases} 0 \text{ if } C_1 = C_2 , \\ n \text{ if } C_1 = C_{i_0} \preceq \ldots \preceq C_{i_n} = C_2 , C_{i_j} \neq C_{i_k} \text{ for } j \neq k , \\ n \text{ if } C_2 = C_{i_0} \preceq \ldots \preceq C_{i_n} = C_1 , C_{i_j} \neq C_{i_k} \text{ for } j \neq k , \\ \infty \text{ otherwise } . \end{cases}$$

For a mapping rule $r = T_j(x) \leftarrow S_i(x)$, $h(r) = T_j$ denotes the target class and $b(r) = S_i$ the source class. Then, these similarity measures are defined:

$$sim(r, r') = \begin{cases} \frac{1}{1+d(b(r), b(r'))} & \text{if } h(r) = h(r') , \\ 0 & \text{otherwise} , \end{cases}$$

$$sim(r, R) = \min_{r' \in R, sim(r,r') > 0} sim(r, r') .$$

These similarities are employed in the definition of hierarchy-based precision and recall:

$$precision = \frac{\sum_{r \in R_L} sim(r, R_A)}{|R_L|}, \quad recall = \frac{\sum_{r \in R_A} sim(r, R_L)}{|R_A|}.$$

Precision measures the average distance of learned rules with the actual ones, while recall measures the average distance of actual rules with the learned ones. Traditional precision and recall are defined in a analogous way, but with equality as similarity measure. In addition, we also combine precision and recall in the F-measure:

$$F = \frac{2}{\frac{1}{precision} + \frac{1}{recall}}.$$

Finally, we also used a variant of traditional precision where we drop all rules for target classes for which there are no relationships at all. This measure shows how good our approach is when we only consider the target classes for which we can be successful.

4.2 Results

In the experiments presented in this section, the learning steps are as follows:

1. Find the best web directory mapping:
   a) Estimate the probabilities $Pr(S_i|T_j, CL_1, \ldots, CL_l)$ for every $S_i \in \mathbf{S}$, $T_j \in \mathbf{T}$ using the web directory-based and text-based classifiers;
   b) Estimate the probabilities $Pr(S_i|T_j)$ for every $S_i \in \mathbf{S}$, $T_j \in \mathbf{T}$ using all classifiers (if any hierarchical classifier is used);
   c) For every target relation T_j and for every non-empty subset of web directory mapping rules having T_j as head, estimate the probability $Pr(\Sigma_j, J, I)$;
   d) Select the rule set Σ_j, which maximizes the probability $Pr(\Sigma_j, J, I)$.
2. Estimate the weights $Pr(T_j|S_i)$ for the learned rules by converting $Pr(S_i|T_j)$, using (3).
3. Compute the error, precision and recall as described above.

The name-based classifiers are slightly modified, as the class names contain some identifier suffix like `Chinese_CHIN_27` (Washington) or `Chinese_30` (Cornell). These suffixes are removed before these classifiers are applied.

The results are depicted in Tables 1–14. Note that they are averaged over both mapping directions Cornell to Washington and vice-versa.

Runs without Constraints

Here, CL_{PN-clm} minimizes the error (0.1305, averaged over all hierarchical classifiers and normalization functions and both mapping directions), the

Table 1. Error

	f_{id}	f_{sum}	$f_{lin} \circ f_{sum}$	$f_{log} \circ f_{sum}$
CL_S	0.2340	0.2439	0.2980	0.3025
CL_{N-clm}	0.2301	0.2932	0.3136	0.2934
CL_{PN-clm}	0.1786	0.1503	0.1515	0.1931
CL_{kNN}	0.2396	0.1971	0.2214	0.2751
CL_B	0.2644	0.1706	0.2252	0.2238
all	0.1139	0.1272	0.1609	0.1611
CL_S / CL_P	0.0778	0.0798	0.0893	0.0892
CL_{N-clm} / CL_P	0.1221	0.1582	0.1616	0.1602
CL_{PN-clm} / CL_P	0.0934	0.1274	0.1211	0.1070
CL_{kNN} / CL_P	0.1098	0.0946	0.1035	0.0943
CL_B / CL_P	0.1167	0.0864	0.1160	0.1190
all / CL_P	0.1087	0.1317	0.1655	0.1443
CL_S / CL_C	0.1554	0.1592	0.1843	0.1864
CL_{N-clm} / CL_C	0.2201	0.2512	0.2333	0.2073
CL_{PN-clm} / CL_C	0.1192	0.1236	0.1228	0.1349
CL_{kNN} / CL_C	0.2332	0.2298	0.2213	0.2791
CL_B / CL_C	0.2771	0.1622	0.2341	0.2413
all / CL_C	0.1195	0.1341	0.1580	0.1622
CL_S / CL_P+CL_C	0.1012	0.1019	0.1109	0.1110
CL_{N-clm} / CL_P+CL_C	0.1472	0.1669	0.1608	0.1528
CL_{PN-clm} / CL_P+CL_C	0.0971	0.1305	0.1246	0.1129
CL_{kNN} / CL_P+CL_C	0.1160	0.1069	0.1106	0.1003
CL_B / CL_P+CL_C	0.1230	0.0927	0.1221	0.1224
all / CL_P+CL_C	0.1159	0.1393	0.1669	0.1480

combination of all classifiers performs about 8% worse. The error of the two content-oriented classifiers–kNN and Naive Bayes–is about 30% worse compared to CL_{PN-clm}, and Naive Bayes is slightly better than kNN. Precision in general is quite low, the highest value is obtained for CL_S (0.2553 on average). The low precision is due to the fact that the system generates a huge number of rules (sometimes several hundreds), while the are only about 50 valid mappings. The best recall is achieved by the combination of all non-hierarchical classifiers with 0.9177 on average. The content-oriented classifiers perform worst, kNN has recall of 0.2772 (nearly 70% worse), while Naive Bayes yields a recall of 0.1920 (about 80% worse). The hierarchical precision and recall values are only slightly better than the traditional ones.

Error and precision is optimized by the f_{sum} normalization function (average error is 0.1524, average precision is 0.0965). The error of the identity function is only slightly worse (1.5%). In addition, this function yields the best recall (0.6322 on average), followed by f_{sum}. Thus, learning parameters for the linear and logistic mapping function does not help.

Table 2. Traditional precision

	f_{id}	f_{sum}	$f_{lin} \circ f_{sum}$	$f_{log} \circ f_{sum}$
CL_S	0.6909	0.6909	0.6909	0.6909
CL_{N-clm}	0.0875	0.0874	0.0875	0.0779
CL_{PN-clm}	0.0339	0.0339	0.0339	0.0541
CL_{kNN}	0.1479	0.1479	0.1479	0.0714
CL_B	0.1525	0.1525	0.1525	0.0939
all	0.0512	0.0685	0.0688	0.0643
CL_S / CL_P	0.0527	0.0525	0.0516	0.0515
CL_{N-clm} / CL_P	0.0680	0.0684	0.0626	0.0549
CL_{PN-clm} / CL_P	0.0360	0.0176	0.0168	0.0088
CL_{kNN} / CL_P	0.0132	0.0191	0.0190	0.0159
CL_B / CL_P	0.0091	0.0127	0.0120	0.0116
all / CL_P	0.0522	0.0740	0.0725	0.0623
CL_S / CL_C	0.2249	0.2255	0.2255	0.2255
CL_{N-clm} / CL_C	0.0912	0.0958	0.0921	0.0815
CL_{PN-clm} / CL_C	0.0356	0.0314	0.0314	0.0448
CL_{kNN} / CL_C	0.0964	0.1235	0.1062	0.0741
CL_B / CL_C	0.0529	0.0993	0.1026	0.0790
all / CL_C	0.0514	0.0686	0.0692	0.0648
CL_S / CL_P+CL_C	0.0535	0.0531	0.0525	0.0526
CL_{N-clm} / CL_P+CL_C	0.0691	0.0698	0.0633	0.0545
CL_{PN-clm} / CL_P+CL_C	0.0368	0.0170	0.0164	0.0098
CL_{kNN} / CL_P+CL_C	0.0152	0.0195	0.0191	0.0165
CL_B / CL_P+CL_C	0.0105	0.0141	0.0137	0.0128
all / CL_P+CL_C	0.0525	0.0741	0.0728	0.0630

Together, the combination of all non-hierarchical classifiers with the identity normalization function yields the lowest error (0.1145 on average), followed by CL_{PN-clm} (about 7% worse). Precision is optimized by using CL_S with any normalization function (virtually the same precision), followed by CL_{N-clm} with f_{sum} or f_{id} (70% worse). Finally, CL_{N-clm} with f_{id} or f_{sum} yields highest recall (9520 and 0.9405, respectively), follows by the classifier combination with the same normalization functions (about 2% worse).

The hierarchical classifier CL_P minimizes error with 0.1157, the combination with CL_C is 7.4% worse. CL_C alone performs more than 60% worse, and using no hierarchical classifier at all increases error by 90% compared to CL_P. This order is reversed for precision and (nearly) recall; best precision is obtained when no hierarchical classifier is used.

Average traditional precision is 0.0919. When only target classes which there are mappings are considered, then precision increases (0.2280 on average).

Table 3. Traditional recall

	f_{id}	f_{sum}	$f_{lin} \circ f_{sum}$	$f_{log} \circ f_{sum}$
CL_S	0.6448	0.6448	0.6448	0.6448
CL_{N-clm}	0.9330	0.9330	0.9330	0.9230
CL_{PN-clm}	0.7730	0.7730	0.7730	0.8459
CL_{kNN}	0.2689	0.2689	0.2689	0.2396
CL_B	0.1241	0.1241	0.1241	0.2196
all	0.9315	0.9415	0.9415	0.8930
CL_S / CL_P	0.7119	0.7119	0.7119	0.7119
CL_{N-clm} / CL_P	0.9615	0.9337	0.8874	0.7841
CL_{PN-clm} / CL_P	0.7922	0.3533	0.3433	0.1633
CL_{kNN} / CL_P	0.2319	0.3152	0.3152	0.2774
CL_B / CL_P	0.1533	0.2004	0.1911	0.1919
all / CL_P	0.9315	0.9322	0.9137	0.8374
CL_S / CL_C	0.7119	0.7119	0.7119	0.7119
CL_{N-clm} / CL_C	0.9522	0.9615	0.9515	0.9322
CL_{PN-clm} / CL_C	0.8015	0.7052	0.7052	0.7681
CL_{kNN} / CL_C	0.2774	0.2689	0.2681	0.2489
CL_B / CL_C	0.1826	0.2096	0.2296	0.2396
all / CL_C	0.9315	0.9415	0.9515	0.9022
CL_S / CL_P+CL_C	0.7219	0.7219	0.7219	0.7219
CL_{N-clm} / CL_P+CL_C	0.9615	0.9337	0.8867	0.7826
CL_{PN-clm} / CL_P+CL_C	0.8015	0.3326	0.3226	0.1833
CL_{kNN} / CL_P+CL_C	0.2596	0.3252	0.3152	0.2867
CL_B / CL_P+CL_C	0.1833	0.2296	0.2311	0.2219
all / CL_P+CL_C	0.9315	0.9322	0.9237	0.8467

Runs with Constraints

In general, precision is quite high, as the number of rules is pruned dramatically. Recall is lower when a constraint is used, but there are often cases where both precision and recall is sufficiently high. However, error is much higher; here, missing rules count much higher than wrong rules with a low weight.

For constraint 1, the differences in the order of the classifiers, normalization functions, the combination of classifier and normalization functions and the order of the hierarchical classifiers are small. In the following, we only present differences to the run without any constraint.

For constraint 2, the differences in the order of the classifiers is neglectable. In contrast, using a linear or logistic normalization function yields better error and precision than the other normalization functions.

The situation is different for constraint 3 ("at most one rule per target rule"), constraint 4 ("only rules with weight above 0.1") and constraint 5 ("at most 40 rules"). Here, CL_S yields the best error and precision; CL_{PN-clm} performs quite bad.

Table 4. Traditional precision, Constraint 3

	f_{id}	f_{sum}	$f_{lin} \circ f_{sum}$	$f_{log} \circ f_{sum}$
CL_S	0.7125	0.7125	0.7008	0.7008
CL_{N-clm}	0.2716	0.2716	0.2716	0.2563
CL_{PN-clm}	0.1283	0.1283	0.1283	0.1443
CL_{kNN}	0.4330	0.4330	0.4330	0.2094
CL_B	0.1681	0.1873	0.1873	0.2265
all	0.2271	0.2352	0.2383	0.2293
CL_S / CL_P	0.2038	0.2038	0.2038	0.2038
CL_{N-clm} / CL_P	0.2205	0.2112	0.2019	0.1863
CL_{PN-clm} / CL_P	0.1365	0.0800	0.0769	0.0241
CL_{kNN} / CL_P	0.0415	0.0484	0.0449	0.0487
CL_B / CL_P	0.0159	0.0399	0.0330	0.0290
all / CL_P	0.2269	0.2322	0.2236	0.2264
CL_S / CL_C	0.5212	0.5125	0.5037	0.5037
CL_{N-clm} / CL_C	0.2869	0.2908	0.2944	0.2942
CL_{PN-clm} / CL_C	0.1394	0.1151	0.1180	0.1277
CL_{kNN} / CL_C	0.4330	0.4523	0.4330	0.2094
CL_B / CL_C	0.0769	0.1695	0.2265	0.2849
all / CL_C	0.2240	0.2414	0.2443	0.2322
CL_S / CL_P+CL_C	0.2069	0.2069	0.2069	0.2069
CL_{N-clm} / CL_P+CL_C	0.2360	0.2267	0.2112	0.1925
CL_{PN-clm} / CL_P+CL_C	0.1423	0.0707	0.0707	0.0272
CL_{kNN} / CL_P+CL_C	0.0415	0.0515	0.0415	0.0487
CL_B / CL_P+CL_C	0.0222	0.0396	0.0321	0.0415
all / CL_P+CL_C	0.2300	0.2355	0.2267	0.2293

Comparison of Constraints

The best error is obtained when no constraint is used at all (0.1622 on average), followed by constraint 1. All other constraints are more than 90% worse; the worst result is obtained when applying all constraints (0.7395). Similarly, recall is maximized when no constraint is used (0.5982 on average), followed by constraint 1. Again, applying all constraints yields the worst recall (0.2906). In contrast, the combination of all constraints yields the highest precision with 0.7584.

The F-measure combines precision and recall. Here, constraint 5 is the best with 0.3745 (where both precision and recall have about the same value), directly followed by the combination of all constraints.

The values for the hierarchical variants are higher (e.g. highest precision with 0.8227), but the order is nearly the same.

The highest hierarchical F-measure 0.6912 (precision of 0.7275, recall of 0.6356) is obtained for CL_S with any of the four normalization functions, without hierarchical classifier, and with constraint 4. The combination of all

Table 5. Traditional recall, Constraint 3

	f_{id}	f_{sum}	$f_{lin} \circ f_{sum}$	$f_{log} \circ f_{sum}$
CL_S	0.5978	0.5978	0.5885	0.5885
CL_{N-clm}	0.6856	0.6856	0.6856	0.6470
CL_{PN-clm}	0.4081	0.4081	0.4081	0.4596
CL_{kNN}	0.2204	0.2204	0.2204	0.1078
CL_B	0.0848	0.0948	0.0948	0.1156
all	0.7263	0.7511	0.7611	0.7326
CL_S / CL_P	0.6078	0.6078	0.6078	0.6078
CL_{N-clm} / CL_P	0.6856	0.6578	0.6307	0.5830
CL_{PN-clm} / CL_P	0.4367	0.2570	0.2470	0.0770
CL_{kNN} / CL_P	0.1278	0.1463	0.1370	0.1456
CL_B / CL_P	0.0493	0.1141	0.0956	0.0878
all / CL_P	0.7256	0.7419	0.7148	0.7233
CL_S / CL_C	0.5978	0.5885	0.5793	0.5793
CL_{N-clm} / CL_C	0.7348	0.7448	0.7541	0.7533
CL_{PN-clm} / CL_C	0.4459	0.3681	0.3774	0.4089
CL_{kNN} / CL_C	0.2204	0.2304	0.2204	0.1078
CL_B / CL_C	0.0400	0.0863	0.1156	0.1463
all / CL_C	0.7163	0.7711	0.7804	0.7419
CL_S / CL_P+CL_C	0.6178	0.6178	0.6178	0.6178
CL_{N-clm} / CL_P+CL_C	0.7348	0.7078	0.6607	0.6030
CL_{PN-clm} / CL_P+CL_C	0.4552	0.2270	0.2270	0.0870
CL_{kNN} / CL_P+CL_C	0.1278	0.1563	0.1278	0.1456
CL_B / CL_P+CL_C	0.0693	0.1148	0.0978	0.1278
all / CL_P+CL_C	0.7356	0.7526	0.7248	0.7326

classifiers with all constraints yield a slightly worse quality for CL_C and the $f_{log} \circ f_{sum}$ normalization function.

Example of Learned Rules

This rule has been learned using the kNN classifier:

```
0.1062 Mechanical_and_Aerospace_Engineering_143(X) :-
           Aeronautics_and_Astronautics_A_A_133(X).
```

Actually, this rule described the only mapping onto the target class `Mechanical_and_Aerospace_Engineering_143`:

```
Mechanical_and_Aerospace_Engineering_143(O) :-
  Mechanical_Engineering_145(O).
```

Table 6. Traditional precision, Constraint 4

	f_{id}	f_{sum}	$f_{lin} \circ f_{sum}$	$f_{log} \circ f_{sum}$
CL_S	0.7063	0.7215	0.7275	0.7275
CL_{N-clm}	0.1052	0.2519	0.4184	0.5342
CL_{PN-clm}	0.0407	0.0490	0.2500	0.0000
CL_{kNN}	0.0000	0.5441	0.4097	0.0848
CL_B	0.0000	0.2833	0.2664	0.1667
all	0.1502	0.5969	0.6892	0.7033
CL_S / CL_P	0.7158	0.7386	0.7326	0.7326
CL_{N-clm} / CL_P	0.1500	0.4319	0.6208	0.6511
CL_{PN-clm} / CL_P	0.0520	0.1623	0.5625	0.6250
CL_{kNN} / CL_P	1.0000	0.5000	1.0000	0.0884
CL_B / CL_P	1.0000	0.3182	0.3500	0.7500
all / CL_P	0.2045	0.6409	0.7766	0.7710
CL_S / CL_C	0.7074	0.7367	0.7305	0.7305
CL_{N-clm} / CL_C	0.1415	0.4342	0.5741	0.6286
CL_{PN-clm} / CL_C	0.0512	0.0833	0.5000	0.1250
CL_{kNN} / CL_C	0.7500	0.4929	0.5000	0.0836
CL_B / CL_C	0.5000	0.2500	0.3167	0.5000
all / CL_C	0.2102	0.6307	0.7636	0.7571
CL_S / CL_P+CL_C	0.7136	0.7305	0.7514	0.7403
CL_{N-clm} / CL_P+CL_C	0.2560	0.5264	0.6186	0.6519
CL_{PN-clm} / CL_P+CL_C	0.0697	0.5833	0.7500	0.4500
CL_{kNN} / CL_P+CL_C	0.7500	0.6667	0.7500	0.6250
CL_B / CL_P+CL_C	0.7500	0.3071	0.4167	0.7500
all / CL_P+CL_C	0.2850	0.6458	0.8001	0.7555

5 Conclusion, Related Work and Outlook

With the proliferation of data sharing applications over the Web, involving ontologies (and in particular web directories), the development of automated tools for ontology matching will be of particular importance. In this paper, we have presented a Probabilistic, Logic-based formal framework (*oPLMap*) for ontology Matching involving web directories. The peculiarity of our approach is that it combines neatly machine learning and heuristic techniques, for learning a set of mapping rules, with logic, in particular probabilistic Datalog. This latter aspect is of particular importance as it constitutes the basis to extend our "ontological "model to more expressive formal languages for ontology description, like OWL DL [21] in particular (founded on so-called *Description Logics* [1]), which are the state of the art in the Semantic Web. As a consequence, all aspects of logical reasoning, considered as important in the Semantic Web community[8], can easily be plugged into our model. Our

[8] http://www.semanticweb.org/

Table 7. Traditional recall, Constraint 4

	f_{id}	f_{sum}	$f_{lin} \circ f_{sum}$	$f_{log} \circ f_{sum}$
CL_S	0.6448	0.6448	0.6356	0.6356
CL_{N-clm}	0.9237	0.7441	0.5507	0.4352
CL_{PN-clm}	0.7452	0.0193	0.0100	0.0100
CL_{kNN}	0.0200	0.2019	0.0678	0.2396
CL_B	0.0000	0.1241	0.0578	0.0093
all	0.8381	0.6578	0.5152	0.5444
CL_S / CL_P	0.6263	0.6170	0.5985	0.5985
CL_{N-clm} / CL_P	0.8774	0.5793	0.4274	0.3681
CL_{PN-clm} / CL_P	0.6989	0.0385	0.0193	0.0293
CL_{kNN} / CL_P	0.0385	0.0778	0.0478	0.1463
CL_B / CL_P	0.0193	0.1041	0.0293	0.0193
all / CL_P	0.8004	0.6378	0.4774	0.4574
CL_S / CL_C	0.6263	0.6170	0.5985	0.5985
CL_{N-clm} / CL_C	0.8681	0.5800	0.3804	0.3596
CL_{PN-clm} / CL_C	0.6896	0.0200	0.0100	0.0200
CL_{kNN} / CL_C	0.0193	0.0885	0.0193	0.1563
CL_B / CL_C	0.0100	0.0763	0.0393	0.0100
all / CL_C	0.8004	0.6193	0.4681	0.4674
CL_S / CL_P+CL_C	0.5985	0.5985	0.5615	0.5330
CL_{N-clm} / CL_P+CL_C	0.7533	0.4930	0.3041	0.3011
CL_{PN-clm} / CL_P+CL_C	0.6619	0.0293	0.0193	0.0293
CL_{kNN} / CL_P+CL_C	0.0193	0.0585	0.0193	0.0478
CL_B / CL_P+CL_C	0.0193	0.0670	0.0193	0.0193
all / CL_P+CL_C	0.7819	0.5722	0.4389	0.4381

logical foundation also eases the formalization of the so-called *query reformulation task* [25], which tackles the issue of converting a query over the target ontology into one (or more) queries over the source ontology.

Our model oPLMap has its foundations in three strictly related research areas: *schema matching* [34], *information integration* [25] and *information exchange* [14], in particular, and borrows from them the terminology and ideas. Indeed, related to the latter, we view the matching problem as the problem of determining the "best possible set Σ of formulae of a certain kind" such that the *exchange* of instances of a source class into a target class has highest probability of being correct. From information integration we inherit the type of rules we are looking for. Indeed, we have a so-called GLaV model [25]. From the former we inherit the requirement to rely on machine learning techniques to automate the process of schema matching. Additionally, a side effect is that we can inherit many of the theoretical results developed in these areas so far, especially from the latter two (see, e.g., [4, 14, 15, 16, 25, 29]).

Table 8. Traditional precision, Constraint 5

	f_{id}	f_{sum}	$f_{lin} \circ f_{sum}$	$f_{log} \circ f_{sum}$
CL_S	0.6909	0.6909	0.6909	0.6909
CL_{N-clm}	0.5600	0.4800	0.4800	0.4800
CL_{PN-clm}	0.3300	0.0400	0.0400	0.0600
CL_{kNN}	0.2700	0.2700	0.2700	0.1100
CL_B	0.1525	0.1525	0.1525	0.1700
all	0.5700	0.6300	0.6600	0.6700
CL_S / CL_P	0.6500	0.6500	0.6400	0.6400
CL_{N-clm} / CL_P	0.5800	0.5100	0.4800	0.4700
CL_{PN-clm} / CL_P	0.3400	0.0800	0.0800	0.0800
CL_{kNN} / CL_P	0.1500	0.1800	0.1600	0.1300
CL_B / CL_P	0.0500	0.1500	0.1100	0.0900
all / CL_P	0.5900	0.6400	0.6700	0.6700
CL_S / CL_C	0.6700	0.6700	0.6700	0.6700
CL_{N-clm} / CL_C	0.5700	0.4800	0.5000	0.5500
CL_{PN-clm} / CL_C	0.3200	0.1400	0.1600	0.1100
CL_{kNN} / CL_C	0.2500	0.2700	0.2800	0.1400
CL_B / CL_C	0.1000	0.1800	0.1800	0.1400
all / CL_C	0.5700	0.6400	0.6700	0.6800
CL_S / CL_P+CL_C	0.6500	0.6400	0.6300	0.6300
CL_{N-clm} / CL_P+CL_C	0.5900	0.5100	0.4800	0.4800
CL_{PN-clm} / CL_P+CL_C	0.3500	0.1200	0.1300	0.1000
CL_{kNN} / CL_P+CL_C	0.1500	0.1700	0.1500	0.1400
CL_B / CL_P+CL_C	0.0800	0.1600	0.1200	0.1100
all / CL_P+CL_C	0.5900	0.6600	0.6700	0.6600

The matching problem for ontologies, as well as the matching problem for schemas has been addressed by many researchers so far and are strictly related, as e.g. schemas can be seen as ontologies with restricted relationship types. The techniques applied in schema matching can be applied to ontology matching as well. Additionally, we have to take care of the hierarchies.

Related to ontology matching are, for instance, the works [10, 22, 24, 32] (see [10] for a more extensive comparison). While most of them use a variety of heuristics to match ontology elements, very few do use machine learning and exploit information in the data instances [10, 22, 24]. [24] computes the similarity between two concepts, as the similarity among the vector representations of the concepts (using Information Retrieval statistics like tf.idf). HICAL [22] uses κ-statistics from the data instances to infer rule mappings among concepts. Finally, [10, 11] (GLUE), but see also [6], is the most involved system. GLUE is based on the ideas introduced earlier by LSD [9]. Similar to our approach, it employed a linear combination of the predictions of multiple base learners (classifiers). The combination weights are learned via

Table 9. Traditional recall, Constraint 5

	f_{id}	f_{sum}	$f_{lin} \circ f_{sum}$	$f_{log} \circ f_{sum}$
CL_S	0.6448	0.6448	0.6448	0.6448
CL_{N-clm}	0.5430	0.4637	0.4637	0.4630
CL_{PN-clm}	0.3204	0.0385	0.0385	0.0578
CL_{kNN}	0.2589	0.2589	0.2589	0.1078
CL_B	0.1241	0.1241	0.1241	0.1626
all	0.5522	0.6078	0.6370	0.6463
CL_S / CL_P	0.6263	0.6263	0.6170	0.6170
CL_{N-clm} / CL_P	0.5615	0.4930	0.4652	0.4552
CL_{PN-clm} / CL_P	0.3304	0.0770	0.0770	0.0770
CL_{kNN} / CL_P	0.1478	0.1756	0.1570	0.1263
CL_B / CL_P	0.0493	0.1433	0.1063	0.0878
all / CL_P	0.5715	0.6170	0.6463	0.6463
CL_S / CL_C	0.6448	0.6448	0.6448	0.6448
CL_{N-clm} / CL_C	0.5530	0.4637	0.4837	0.5300
CL_{PN-clm} / CL_C	0.3111	0.1356	0.1556	0.1070
CL_{kNN} / CL_C	0.2404	0.2589	0.2681	0.1370
CL_B / CL_C	0.0978	0.1726	0.1726	0.1356
all / CL_C	0.5522	0.6178	0.6463	0.6556
CL_S / CL_P+CL_C	0.6263	0.6170	0.6078	0.6078
CL_{N-clm} / CL_P+CL_C	0.5715	0.4930	0.4659	0.4659
CL_{PN-clm} / CL_P+CL_C	0.3396	0.1170	0.1270	0.0970
CL_{kNN} / CL_P+CL_C	0.1478	0.1663	0.1478	0.1363
CL_B / CL_P+CL_C	0.0793	0.1533	0.1178	0.1078
all / CL_P+CL_C	0.5715	0.6363	0.6463	0.6370

regression on manually specified mappings between a small number of learning ontologies.

Related to schema matching are, for instance, the works [3, 6, 7, 8, 9, 13, 14, 18, 19, 23, 27, 28, 30, 33, 36] (see [34] for a more extensive comparison). As pointed out above, closest to our approach is [14] based on a logical framework for data exchange, but we incorporated the inherent uncertainty of rule mappings and classifier combinations (like LSD) into our framework as well. While the majority of the approaches focuses on finding 1-1 matchings (e.g. iMap [6] is an exception), we allow complex mappings and domain knowledge as well.

As future work, we see some appealing points. The combination of a rule-based language with an expressive ontology language has attracted the attention of many researchers (see, e.g., [12, 20] to cite a few) and is considered as an important requirement. Currently we are combining probabilistic Datalog with OWL DL so that complex ontologies can be described (so far, none of the approaches above addresses the issue of uncertainty). Besides this, as then

Table 10. Traditional precision, All constraints

	f_{id}	f_{sum}	$f_{lin} \circ f_{sum}$	$f_{log} \circ f_{sum}$
CL_S	0.7790	0.7790	0.7790	0.7596
CL_{N-clm}	0.5600	0.4900	0.4928	0.5546
CL_{PN-clm}	0.3400	0.0714	0.2500	0.0000
CL_{kNN}	0.0000	0.4722	0.3250	0.1664
CL_B	0.0000	0.2798	0.2976	0.1667
all	0.6100	0.6625	0.7393	0.7421
CL_S / CL_P	0.7625	0.7732	0.7867	0.7663
CL_{N-clm} / CL_P	0.5800	0.5100	0.6934	0.7054
CL_{PN-clm} / CL_P	0.3400	0.2917	0.6250	0.7500
CL_{kNN} / CL_P	1.0000	0.4250	1.0000	0.2929
CL_B / CL_P	1.0000	0.3289	0.3500	0.7500
all / CL_P	0.6200	0.6951	0.7999	0.7952
CL_S / CL_C	0.7829	0.7936	0.7715	0.7715
CL_{N-clm} / CL_C	0.5700	0.4841	0.5872	0.6148
CL_{PN-clm} / CL_C	0.3800	0.0714	0.5000	0.0833
CL_{kNN} / CL_C	0.5000	0.5000	0.2500	0.1692
CL_B / CL_C	0.5000	0.2938	0.1500	0.5000
all / CL_C	0.6000	0.6970	0.7999	0.7837
CL_S / CL_P+CL_C	0.7887	0.7998	0.8117	0.8099
CL_{N-clm} / CL_P+CL_C	0.5900	0.5881	0.6952	0.7396
CL_{PN-clm} / CL_P+CL_C	0.4000	1.0000	1.0000	1.0000
CL_{kNN} / CL_P+CL_C	1.0000	0.7500	1.0000	0.7500
CL_B / CL_P+CL_C	1.0000	0.3654	0.6667	1.0000
all / CL_P+CL_C	0.6300	0.7196	0.8090	0.7992

the instances of a class may be structured, e.g. have several attributes, or may be semi-structured (e.g. XML documents) we have to combine our ontology matching method with a so-called schema matching method (see, e.g. [9, 34]). We plan to integrate our model with the method based on [31] for schema matching, as the latter is rooted on the same principles of the work we have presented here. In particular, we are investigating several methods to learn mappings in an environment with ontologies and structured or semi-structured data: (*i*) to learn the schema mappings for each ontology class first, and ontology mappings in a second step; or (*ii*) both learning steps are performed simultaneously, which means that the quality of every possible mapping rule is estimated, and an overall optimum mapping subset is selected.

Additional areas of intervention rely on augmenting the effectiveness of the machine learning part. While to fit new classifiers into our model is straightforward theoretically, practically finding out the most appropriate one or a combination of them is quite more difficult, as our results show. In the future, more variants should be developed and evaluated to improve the quality of

Table 11. Traditional recall, All constraints

	f_{id}	f_{sum}	$f_{lin} \circ f_{sum}$	$f_{log} \circ f_{sum}$
CL_S	0.5785	0.5785	0.5785	0.5785
CL_{N-clm}	0.5430	0.4744	0.4652	0.4259
CL_{PN-clm}	0.3304	0.0100	0.0100	0.0100
CL_{kNN}	0.0200	0.0870	0.0285	0.0778
CL_B	0.0000	0.0763	0.0578	0.0093
all	0.5907	0.5815	0.4959	0.4959
CL_S / CL_P	0.5600	0.5600	0.5415	0.5415
CL_{N-clm} / CL_P	0.5622	0.4944	0.4181	0.3589
CL_{PN-clm} / CL_P	0.3296	0.0385	0.0193	0.0193
CL_{kNN} / CL_P	0.0385	0.0378	0.0478	0.1170
CL_B / CL_P	0.0193	0.0856	0.0293	0.0193
all / CL_P	0.6000	0.5707	0.4581	0.4381
CL_S / CL_C	0.5600	0.5600	0.5322	0.5322
CL_{N-clm} / CL_C	0.5530	0.4652	0.3611	0.3404
CL_{PN-clm} / CL_C	0.3696	0.0100	0.0100	0.0100
CL_{kNN} / CL_C	0.0100	0.0393	0.0100	0.0685
CL_B / CL_C	0.0100	0.0670	0.0300	0.0100
all / CL_C	0.5807	0.5622	0.4581	0.4381
CL_S / CL_P+CL_C	0.5515	0.5515	0.5237	0.5137
CL_{N-clm} / CL_P+CL_C	0.5722	0.4652	0.2948	0.2919
CL_{PN-clm} / CL_P+CL_C	0.3889	0.0293	0.0193	0.0293
CL_{kNN} / CL_P+CL_C	0.0193	0.0285	0.0193	0.0478
CL_B / CL_P+CL_C	0.0193	0.0670	0.0193	0.0193
all / CL_P+CL_C	0.6100	0.5344	0.4289	0.4189

the learning mechanism. Additional classifiers could consider the data types of two classes, could use a thesaurus for finding synonym class names, or could use other measures like KL-distance or mutual information (joint entropy). Furthermore, instead of averaging the classifier predictions, the weights of each classifier could be learned via regression. Another interesting direction of investigation would be to evaluate the effect to integrate our model with graph-matching algorithms like, for instance [23, 30]. These can be considered as additional classifiers. Last, but not least it would be interesting to evaluate the effect of allowing more expressive mapping rules, as for instance of the form $\alpha_{j,i} T_j(x) \leftarrow S_{i_1}(x), \ldots, S_{i_n}(\mathrm{x})$ or more generally on full featured logic programming or of the form presented in [5], as well as to consider the impact of probabilities $Pr(\bar{S}_i|T_j)$, $Pr(\bar{S}_i|\bar{T}_j)$, $Pr(S_i|\bar{T}_j)$[9] (and vice-versa inverting T_j with S_i).

[9] $\bar{X}$ is the complement of X.

Table 12. Hierarchical precision, All constraints

	f_{id}	f_{sum}	$f_{lin} \circ f_{sum}$	$f_{log} \circ f_{sum}$
CL_S	0.8053	0.8053	0.8053	0.7859
CL_{N-clm}	0.5800	0.5100	0.5087	0.5615
CL_{PN-clm}	0.3683	0.1409	0.5000	0.0714
CL_{kNN}	0.5000	0.6241	0.5833	0.2547
CL_B	0.0000	0.3562	0.3704	0.1667
all	0.6433	0.6958	0.7571	0.7514
CL_S / CL_P	0.7903	0.7946	0.8094	0.7890
CL_{N-clm} / CL_P	0.6000	0.5300	0.6934	0.7054
CL_{PN-clm} / CL_P	0.3733	0.2917	0.6250	0.7500
CL_{kNN} / CL_P	1.0000	0.6333	1.0000	0.3249
CL_B / CL_P	1.0000	0.3980	0.4167	0.7500
all / CL_P	0.6533	0.7241	0.8118	0.8071
CL_S / CL_C	0.8107	0.8150	0.8018	0.8018
CL_{N-clm} / CL_C	0.5950	0.5096	0.5972	0.6248
CL_{PN-clm} / CL_C	0.4133	0.3214	0.7500	0.3333
CL_{kNN} / CL_C	0.7500	0.7292	0.6250	0.2588
CL_B / CL_C	0.7500	0.4062	0.3250	0.6250
all / CL_C	0.6333	0.7237	0.8118	0.7956
CL_S / CL_P+CL_C	0.8181	0.8226	0.8376	0.8357
CL_{N-clm} / CL_P+CL_C	0.6100	0.6000	0.6952	0.7396
CL_{PN-clm} / CL_P+CL_C	0.4283	1.0000	1.0000	1.0000
CL_{kNN} / CL_P+CL_C	1.0000	0.8750	1.0000	0.7500
CL_B / CL_P+CL_C	1.0000	0.4351	0.6667	1.0000
all / CL_P+CL_C	0.6633	0.7417	0.8090	0.8117

References

1. Franz Baader, Diego Calvanese, Deborah McGuinness, Daniele Nardi, and Peter F. Patel-Schneider, editors. *The Description Logic Handbook: Theory, Implementation, and Applications.* Cambridge University Press, 2003.
2. Ricardo A. Baeza-Yates and Berthier Ribeiro-Neto. *Modern Information Retrieval.* Addison-Wesley Longman Publishing Co., Inc., 1999.
3. J. Berlin and A. Motro. Database schema matching using machine learning with feature selection. In *In Proceedings of the Conf. on Advanced Information Systems Engineering (CAiSE), 2002.*, 2002.
4. Diego Calvanese, Giuseppe De Giacomo, Maurizio Lenzerini, and Moshe Y. Vardi. Lossless regular views. In *Proc. of the 21st ACM SIGACT SIGMOD SIGART Sym. on Principles of Database Systems (PODS 2002)*, pages 247–258, 2002.
5. Hans Chalupsky. Ontomorph: a translation system for symbolic knowledge. In *Proceedings of the 9th International Conference on Principles of Knowledge Representation and Reasoning (KR-04).* AAAI Press, 2000.

Table 13. Hierarchical recall, All constraints

	f_{id}	f_{sum}	$f_{lin} \circ f_{sum}$	$f_{log} \circ f_{sum}$
CL_S	0.5970	0.5970	0.5970	0.5970
CL_{N-clm}	0.5615	0.4930	0.4791	0.4306
CL_{PN-clm}	0.3573	0.0177	0.0146	0.0100
CL_{kNN}	0.0200	0.1140	0.0509	0.1194
CL_B	0.0000	0.1043	0.0757	0.0093
all	0.6220	0.6083	0.5052	0.5006
CL_S / CL_P	0.5785	0.5739	0.5554	0.5554
CL_{N-clm} / CL_P	0.5807	0.5130	0.4181	0.3589
CL_{PN-clm} / CL_P	0.3616	0.0385	0.0193	0.0193
CL_{kNN} / CL_P	0.0385	0.0555	0.0478	0.1301
CL_B / CL_P	0.0193	0.1085	0.0426	0.0193
all / CL_P	0.6312	0.5930	0.4628	0.4428
CL_S / CL_C	0.5785	0.5739	0.5507	0.5507
CL_{N-clm} / CL_C	0.5761	0.4883	0.3657	0.3450
CL_{PN-clm} / CL_C	0.4012	0.0146	0.0146	0.0146
CL_{kNN} / CL_C	0.0146	0.0566	0.0196	0.1020
CL_B / CL_C	0.0146	0.0963	0.0446	0.0146
all / CL_C	0.6120	0.5811	0.4628	0.4428
CL_S / CL_P+CL_C	0.5700	0.5654	0.5376	0.5276
CL_{N-clm} / CL_P+CL_C	0.5907	0.4744	0.2948	0.2919
CL_{PN-clm} / CL_P+CL_C	0.4159	0.0293	0.0193	0.0293
CL_{kNN} / CL_P+CL_C	0.0193	0.0335	0.0193	0.0478
CL_B / CL_P+CL_C	0.0193	0.0817	0.0193	0.0193
all / CL_P+CL_C	0.6412	0.5487	0.4289	0.4235

Table 14. Overall traditional precision, recall and F-Measure

	Precision	Recall	F-Measure
No constraint	0.0919	0.5982	0.1307
Constraint 1	0.0939	0.5970	0.1323
Constraint 2	0.1495	0.4677	0.1856
Constraint 3	0.2185	0.4365	0.2621
Constraint 4	0.4995	0.3451	0.3012
Constraint 5	0.3821	0.3676	0.3745
All constraints	0.5979	0.2819	0.3284

6. Robin Dhamankar, Yoonkyong Lee, AnHai Doan, Alon Halevy, and Pedro Domingos. iMAP: discovering complex semantic matches between database schemas. In *Proceedings of the 2004 ACM SIGMOD international conference on Management of data*, pages 383–394. ACM Press, 2004.
7. H. Do and E. Rahm. Coma - a system for flexible combination of schema matching approaches. In *Proceedings of the Int. Conf. on Very Large Data Bases (VLDB-02), 2002.*, 2002.
8. Anhai Doan, Pedro Domingos, and Alon Halevy. Learning to match the schemas of data sources: A multistrategy approach. *Mach. Learn.*, 50(3):279–301, 2003.
9. AnHai Doan, Pedro Domingos, and Alon Y. Halevy. Reconciling schemas of disparate data sources: a machine-learning approach. In *Proceedings of the 2001 ACM SIGMOD international conference on Management of data*, pages 509–520. ACM Press, 2001.
10. AnHai Doan, Jayant Madhavan, Robin Dhamankar, Pedro Domingos, and Alon Halevy. Learning to match ontologies on the semantic web. *The VLDB Journal*, 12(4):303–319, 2003.
11. AnHai Doan, Jayant Madhavan, Pedro Domingos, and Alon Halevy. Learning to map between ontologies on the semantic web. In *Proceedings of the eleventh international conference on World Wide Web*, pages 662–673. ACM Press, 2002.
12. Thomas Eiter, Thomas Lukasiewicz, Roman Schindlauer, and Hans Tompits. Combining answer set programming with description logics for the semantic web. In *Proceedings of the 9th International Conference on Principles of Knowledge Representation and Reasoning (KR-04)*. AAAI Press, 2004.
13. David W. Embley, David Jackman, and Li Xu. Multifaceted exploitation of metadata for attribute match discovery in information integration. In *Workshop on Information Integration on the Web*, pages 110–117, 2001.
14. Ronald Fagin, Phokion G. Kolaitis, Reneé Miller, and Lucian Popa. Data exchange: Semantics and query answering. In *Proceedings of the International Conference on Database Theory (ICDT-03)*, number 2572 in Lecture Notes in Computer Science, pages 207–224. Springer Verlag, 2003.
15. Ronald Fagin, Phokion G. Kolaitis, and Lucian Popa. Data exchange: getting to the core. In *Proceedings of the twenty-second ACM SIGMOD-SIGACT-SIGART symposium on Principles of database systems*, pages 90–101. ACM Press, 2003.
16. Ronald Fagin, Phokion G. Kolaitis, Wang-Chiew Tan, and Lucian Popa. Composing schema mappings: Second-order dependencies to the rescue. In *Proceedings PODS*, 2004.
17. Norbert Fuhr. Probabilistic Datalog: Implementing logical information retrieval for advanced applications. *Journal of the American Society for Information Science*, 51(2):95–110, 2000.
18. MingChuan Guo and Yong Yu. Mutual enhancement of schema mapping and data mapping. In *In ACM SIGKDD 2004 Workshop on Mining for and from the Semantic Web*, Seattle, 2004.
19. Bin He and Kevin Chen-Chuan Chang. Statistical schema matching across web query interfaces. In *Proceedings of the 2003 ACM SIGMOD international conference on Management of data*, pages 217–228. ACM Press, 2003.

20. Ian Horrocks and Peter F. Patel-Schneider. A proposal for an OWL rules language. In *Proc. of the Thirteenth International World Wide Web Conference (WWW-04)*. ACM, 2004.
21. Ian Horrocks, Peter F. Patel-Schneider, and Frank van Harmelen. From SHIQ and RDF to OWL: The making of a web ontology language. *Journal of Web Semantics*, 1(1):7–26, 2003.
22. R. Ichise, H. Takeda, and S. Honiden. Rule induction for concept hierarchy alignment. In *Proceedings of the Workshop on Ontology Learning at the 17th International Joint Conference on Artificial Intelligence (IJCAI-01)*, 2001.
23. Jaewoo Kang and Jeffrey F. Naughton. On schema matching with opaque column names and data values. In *Proceedings of the 2003 ACM SIGMOD international conference on on Management of data*, pages 205–216. ACM Press, 2003.
24. Martin S. Lacher and Georg Groh. Facilitating the exchange of explicit knowledge through ontology mappings. In *Proceedings of the Fourteenth International Florida Artificial Intelligence Research Society Conference*, pages 305–309. AAAI Press, 2001.
25. Maurizio Lenzerini. Data integration: a theoretical perspective. In *Proceedings of the 21st ACM SIGMOD-SIGACT-SIGART symposium on Principles of database systems (PODS-02)*, pages 233–246. ACM Press, 2002.
26. John W. Lloyd. *Foundations of Logic Programming.* Springer, Heidelberg, RG, 1987.
27. J. Madhavan, P. Bernstein, K. Chen, A. Halevy, and P. Shenoy. Corpus-based schema matching. In *Workshop on Information Integration on the Web at IJCAI-03*, 2003.
28. Jayant Madhavan, Philip A. Bernstein, and Erhard Rahm. Generic schema matching with cupid. In *Proc. 27th VLDB Conference*, pages 49–58, 2001.
29. Ronald Fagin Marcelo Arenas, Pablo Barcelo and Leonid Libkin. Locally consistent transformations and query answering in data exchange. In *Proceedings of the 23st ACM SIGMOD-SIGACT-SIGART symposium on Principles of database systems (PODS-04)*, pages 229–240. ACM Press, 2004.
30. S. Melnik, H. Garcia-Molina, and E. Rahm. Similarity flooding: A versatile graph matching algorithm and its application to schema matching. In *Proceedings of the 18th International Conference on Data Engineering (ICDE'02)*, page 117. IEEE Computer Society, 2002.
31. Henrik Nottelmann and Umberto Straccia. A probabilistic approach to schema matching. Technical Report 2004-TR-60, Istituto di Scienza e Tecnologie dell'Informazione, Consiglio Nazionale delle Ricerche, Pisa, Italy, 2004.
32. Natalya Fridman Noy and Mark A. Musen. Prompt: Algorithm and tool for automated ontology merging and alignment. In *Proceedings of the Seventeenth National Conference on Artificial Intelligence and Twelfth Conference on Innovative Applications of Artificial Intelligence*, pages 450–455. AAAI Press / The MIT Press, 2000.
33. Lucian Popa, Yannis Velegrakis, Renee J. Miller, Mauricio A. Hernandez, and Ronald Fagin. Translating web data. In *Proceedings of VLDB 2002, Hong Kong SAR, China*, pages 598–609, 2002.
34. Erhard Rahm and Philip A. Bernstein. A survey of approaches to automatic schema matching. *The VLDB Journal*, 10(4):334–350, 2001.

35. Fabrizio Sebastiani. Machine learning in automated text categorization. *ACM Computing Surveys*, 34(1):1–47, 2002.
36. Ling Ling Yan, Renée J. Miller, Laura M. Haas, and Ronald Fagin. Data-driven understanding and refinement of schema mappings. In *Proceedings of the 2001 ACM SIGMOD international conference on Management of data*, pages 485–496. ACM Press, 2001.

The SP Theory and the Representation and Processing of Knowledge

J. Gerard Wolff

CognitionResearch.org.uk, Menai Bridge, UK.
jgw@cognitionresearch.org.uk

Summary. This chapter describes an approach to the representation and processing of knowledge, based on the SP theory of computing and cognition. This approach has strengths that complement others such as those currently proposed for the Semantic Web. The benefits of the SP approach are simplicity and comprehensibility in the representation of knowledge, an ability to cope with errors and uncertainties in knowledge, and capabilities for 'intelligent' processing of knowledge, including probabilistic reasoning, pattern recognition, information retrieval, unsupervised learning, planning and problem solving.

Key words: information compression, multiple alignment, semantic web, ontologies, probabilistic reasoning, pattern recognition, information retrieval, unsupervised learning, planning and problem solving.

1 Introduction

The SP theory is a new theory of computing and cognition that integrates and simplifies concepts in those fields (see [27] and earlier publications cited there). The purpose of this chapter is to describe how the SP theory may be applied to the representation and processing of knowledge and to compare it with some of the alternatives, such those currently proposed for the Semantic Web [2].

The main benefits of the SP approach are:

- Simplicity, comprehensibility and versatility in the representation of knowledge.
- An ability to cope with errors and uncertainties in knowledge.
- Capabilities for 'intelligent' processing of knowledge, including probabilistic reasoning, fuzzy pattern recognition, information retrieval, unsupervised learning, planning and problem solving.

The next section provides a brief introduction to the SP theory. Then a preliminary example is presented showing how knowledge can be expressed in the framework and how it can be processed by building 'multiple alignments'. Sect. 3 presents a more elaborate example showing how class hierarchies and part-whole hierarchies

J.G. Wolff: *The SP Theory and the Representation and Processing of Knowledge*, StudFuzz **204**, 79–101 (2006)
www.springerlink.com

may be integrated and processed within the SP system. This section also discusses more generally how a range of constructs may be expressed. Sect. 4 reviews the capabilities of the SP system in other areas of AI and Sect. 5 compares the SP approach to the representation and processing of knowledge with alternatives that are being developed for the Semantic Web.

2 The SP Theory

The SP theory grew out of a long tradition in psychology that many aspects of perception, cognition and the workings of brains and nervous systems may be understood as information compression. It is founded on principles of 'minimum length encoding', pioneered by [14, 17, 18] and others (see [11]).

The theory is conceived as an abstract model of *any* system for processing information, either natural or artificial. In broad terms, it receives data (designated 'New') from its environment and adds these data to a body of stored of knowledge (called 'Old'). At the same time, it tries to compress the information as much as possible by searching for full or partial matches between patterns and unifying patterns or subpatterns that are the same. In the process of compressing information, the system builds *multiple alignments* of the kind shown below.

In the SP system *all* kinds of knowledge are represented by arrays of atomic symbols in one or two dimensions called 'patterns'. Despite the extreme simplicity of this format, the way these 'flat' patterns are processed in the SP system means that they can be used to model a variety of established schemes including class-hierarchies and part-whole hierarchies (as we shall see below), discrimination networks and trees, condition-action rules, context-free and context-sensitive grammars and others. The provision of a uniform format for all these kinds of knowledge facilitates their seamless integration.

The SP system is Turing-equivalent in the sense that it can model the operation of a universal Turing machine [21] but it is built from different foundations and it provides mechanisms—not provided in a 'raw' Turing machine—for the matching and unification of patterns and the building multiple alignments. These mechanisms facilitate the integration and simplification of a range of operations, especially in artificial intelligence.

To date, the main areas in which the SP framework has been applied are probabilistic reasoning, pattern recognition and information retrieval [22], parsing and production of natural language [23], modelling concepts in logic and mathematics [25], and unsupervised learning [26, 28].

2.1 Computer Models

Two computer models of the SP system have been developed:

- The SP62 model is a partial realisation of the framework that builds multiple alignments but does not add to the store of Old knowledge. It also calculates probabilities for inferences that can be drawn from the multiple alignments. This model, which is relatively robust and mature, is an enhanced version of the SP61 model described in [23]. Most of the examples presented in this chapter are output from SP62.[1]

- The SP70 model embodies all elements of the framework including the process of building the repository of Old, stored knowledge. This model already has a capability for the unsupervised learning of grammars and similar kinds of knowledge but further work is needed to realise the full potential of the model in this area. A description of the model and its capabilities may be found in [26, 28].

2.2 Introductory Example

To introduce the multiple alignment concept as it has been developed in this research, this section presents an example showing how knowledge can be represented in the SP framework and how it can be processed with the SP62 model. This first example is intended to suggest, in a preliminary way, how ontologies (in the Semantic Web or elsewhere) may be represented and processed in the SP framework.

The example presented here and examples in the rest of the chapter are fairly small. This is partly to save space but mainly because small examples are easier to understand than big ones. The relatively small size of the examples should not be taken to represent the limits of what can be done with the model. More complex examples may be seen in [23].

Representing Knowledge with Patterns

Consider, first of all, the set of SP 'patterns' shown in Fig. 1. In the first row, the pattern '<animal> eats breathes has-senses ... </animal>' describes the main attributes of the class 'animal', including such features as eating, breathing and being sensitive to stimulation. In a more comprehensive description of the class, the three dots ('...') would be replaced by other symbols such as 'reproduction' (animals produce offspring), 'locomotion' (animals can normally move from place to place), and so on.

In a similar way, the second pattern in the figure ('<mammal> <animal> </animal> furry warm-blooded ... </mammal>') describes the class 'mammal', the third pattern describes the class 'cat', while the fourth pattern describes a particular cat ('Tibs') and its individual attributes. The patterns that follow describe the classes 'reptile', 'bird', 'robin' and an individual called 'Tweety'.

Notice that the pattern that describes the class 'mammal' contains the pair of symbols '<animal> </animal>'. As we shall see, this pair of symbols serves to show that mammals belong in the class of animals and have the attributes of animals. In a similar way, the pair of symbols '<mammal> </mammal>' within the

[1] The main difference between SP62 and SP61 is that the former allows one *or more* New patterns in each multiple alignment whereas SP61 allows only one. The source code and a Windows executable for SP62 may be obtained from `www.cognitionresearch.org.uk/sp.htm`.

```
<animal> eats breathes has-senses ... </animal>
<mammal> <animal> </animal> furry warm-blooded ... </mammal>
<cat> <mammal> </mammal> purrs retractile-claws ... </cat>
<Tibs> <cat> </cat> tabby <chest> white </chest> ... </Tibs>
<reptile> <animal> </animal> cold-blooded scaly-skin ... </reptile>
<bird> <animal> </animal> wings feathers can-fly ... </bird>
<robin> <bird> </bird> red-breast ... </robin>
<Tweety> <robin> </robin> ... </Tweety>
```

Fig. 1. A set of SP patterns describing various kinds of animal at varying levels of abstraction

pattern that describes the class of cats serves to show that cats are mammals with the characteristics of mammals, and likewise with other patterns in Fig. 1

Basic Concepts

Before we proceed, let us briefly review the main constructs that we have seen so far.

As previously noted, an SP *pattern* is an array of 'symbols' in one or two dimensions. In work to date, the main focus has been on one-dimensional patterns but it is envisaged that, at some stage, the concepts will be generalised for patterns in two dimensions. These would provide a natural vehicle for representing such things as maps, diagrams or pictures.

An SP *symbol* is a sequence of one or more non-space characters bounded by white space. It is simply a 'mark' that can be matched in an all-or-nothing manner with other symbols.

Unlike symbols in systems like arithmetic, SP symbols have no intrinsic meaning such as 'add' for the symbol '+' or 'multiply' for the symbol '×'. Any meaning that attaches to an SP symbol must be expressed as one or more other SP symbols that are associated with the given symbol in a given set of patterns.

Within the SP framework, these simple constructs provide a powerful means for the representation of diverse kinds of knowledge and their seamless integration (see also Sect. 4.1, below).

Boundary Markers

Readers will notice that each of the patterns in Fig. 1 starts with a symbol like '<animal>' and ends with a corresponding symbol like '</animal>', much like the start tags and end tags used in HTML (www.w3.org/TR/html4/), XML (www.w3.org/XML/) or RDF (www.w3.org/RDF/). However, by contrast with start tags and end tags in those languages, symbols like '<animal>' and '</animal>' have no formal status in the SP system. Any convenient style may be used to mark the beginnings and ends of patterns, such as 'animal ... #animal' or 'animal ... %animal'. It also possible to use the left bracket, '<', to mark the start of a pattern with the corresponding right bracket, '>', at the end of the pattern (examples will be seen in Sect. 4.4, below). As we shall see, boundary markers are not always required for every pattern and in some applications they may not be needed in any patterns.

Building Multiple Alignments

If SP62 is run with the patterns from Fig. 1 in its repository of Old information, and with the set of patterns {'furry', '<chest> white </chest>', 'playful', 'purrs', 'eats'} as its New information, the program forms several multiple alignments, the best one of which is shown in Fig. 2. The meaning of 'best' in this context is described below.

```
0          1          2                  3              4

           <Tibs>
           <cat> ---- <cat>
                      <mammal> --------- <mammal>
                                         <animal> ----- <animal>
eats ---------------------------------------------- eats
                                                        breathes
                                                        has-senses
                                                        ...
                                         </animal> ---- </animal>
furry ---------------------------------- furry
                                         warm-blooded
                                         ...
                      </mammal> -------- </mammal>
purrs -------------- purrs
                      retractile-claws
                      ...
           </cat> --- </cat>
           tabby
<chest> -- <chest>
white ---- white
</chest> - </chest>
playful    ...
           </Tibs>

0          1          2                  3              4
```

Fig. 2. The best multiple alignment found by SP62 with the patterns from Fig. 1 in Old and the set of one-symbol patterns {'furry', '<chest> white </chest>', 'playful', 'purrs', 'eats'} in New

In this and other multiple alignments to be shown below, column 0 contains one or more New patterns while each of the remaining columns contains one Old pattern and only one such pattern. The order of the Old patterns in columns to the right of column 0 is entirely arbitrary and without special significance. In the multiple alignments, symbols that match each other from one pattern to another are connected by broken lines.

The patterns are arranged vertically in these examples because this allows the alignments to fit better on the page. In other alignments shown later, the patterns are arranged horizontally. In this case, the New pattern or patterns are always in the top row with the Old patterns in rows underneath. As with the vertical arrangement, the order of the Old patterns in the multiple alignment has no special significance.

Notice that the order of the New patterns in column 0 may be different from their order as they were supplied to the program. However, within any pattern containing two or more symbols (such as '<chest> white </chest>' in our example), the order of the symbols must be preserved.

Notice also that it is not necessary for every New symbol to be matched with an Old symbol: in this example, 'playful' is not matched with any other symbol. Likewise, it is not necessary for every Old symbol to be matched with any other symbol.

In what sense is this multiple alignment the 'best' of the multiple alignments formed by SP62? It is best because it has the highest 'compression score' calculated by SP62. This score is a measure of the amount of compression of the New pattern or patterns that can be achieved by encoding those patterns in terms of the Old patterns in the multiple alignment. The details of how this encoding is done and how the score is calculated are explained in Appendix B of [23].

2.3 Interpretation: Recognition, Retrieval and Reasoning

How should we interpret a multiple alignment like the one shown in Fig. 2? The most natural interpretation is that it represents the result of a process of recognition. An unknown entity, with the features {'furry', '<chest> white </chest>', 'purrs', 'eats'}, has been recognised as being the entity 'Tibs' (column 1). At the same time, it is recognised as being a cat (column 2), a mammal (column 3) and an animal (column 4).

The formation of a multiple alignment like this may also be interpreted as a process of information retrieval. The information in New may be viewed as a 'query' applied to a database of patterns, somewhat in the manner of 'query-by-example', and the patterns in the best multiple alignment may be seen as an answer to the query.

A major benefit of this process of recognition-cum-retrieval is that it allows us to make several inferences about the unknown entity. Although we may have had just a glimpse of Tibs that allowed us to see only his white chest, we can infer from the details in column 1 that, apart from his chest, Tibs has a tabby colouration. Likewise, we can infer that, as a cat, Tibs has retractile claws and other attributes of cats (column 2), as a mammal he is warm blooded, furry, and so on (column 3), and that, as an animal, Tibs breathes, is sensitive to stimulation, and so on (column 4). In short, the formation of a multiple alignment like the one in Fig. 2 provides for the recognition of an entity with *inheritance* of attributes through several levels of abstraction, in the manner of object-oriented design.

The key idea is that any Old symbol within a multiple alignment that is not matched with a New symbol represents an inference that may be drawn from the alignment. This is true, regardless of the size or complexity of the multiple alignment. As we shall see in Sect. 4.3, below, SP62 allows us to calculate absolute and relative probabilities for multiple alignments and the inferences that they represent.

2.4 User Interface

There is no suggestion here that multiple alignments like the one shown in Fig. 2 would necessarily be presented to users for inspection. It is possible that users might find it useful to see multiple alignments but it seems more likely that alignments would be formed 'behind the scenes' and users would see the results presented in some other format.

The formation of multiple alignments is, primarily, a computational technique for the recognition or retrieval of patterns and the drawing of probabilistic inferences. How the results should best be presented to users is a question that is outside the scope of the present chapter.

3 Recognition, Retrieval and Reasoning with Part-Whole Relations and Class-Inclusion Relations

This section presents a slightly more elaborate example designed to show how, within the SP framework, a class-inclusion hierarchy may be integrated with a part-whole hierarchy. It also provides a simple example of the way in which reasoning may be integrated with recognition and retrieval.

Figure 3 shows a set of patterns which are a partial description of the class 'person' and associated concepts. The first pattern in the figure describes the class person in broad-brush terms. In this description, a pair of symbols like '<first-name> </first-name>' represents a 'property' or 'variable' without any assigned 'value'. In this description, a person has a first name, last name, gender and profession (all with unspecified values) and it also has the parts 'head', 'body' and 'legs'. The pattern also shows a variable for the kind of voice that a given person has. As before, three dots ('...') are used to represent other attributes that would be included in a fuller description.

The next three patterns show well-known associations between common first names and the gender of the person who has that name. Anyone called 'Mary' is very likely to be female whereas someone called 'Jack' or 'Peter' is almost certainly male.

The two patterns beginning with the symbols '<person> <gender> ...' provide partial descriptions of each gender, including descriptions relating to a person's chin and the nature of their voice. In this context, the symbol 'beard' is intended as a shorthand for the idea that a male person may grow a beard although any given male person might be clean-shaven. No attempt has been made in this simple example to describe children or any other exceptions to the rule.

Each of the four patterns for 'profession' provide a representative attribute in each case: a doctor has a stethoscope, a lawyer has law books, and so on.

The pattern which follows—'<head> <hair> </hair> <eyes> </eyes> <nose> </nose> <mouth> </mouth> <chin> </chin> ... </head>'—describes the structure of a person's head and the pattern after that provides two variables for a person's hair: its colour and its length. In a fuller description of the class 'person',

```
<person> <first-name> </first-name> <last-name> </last-name>
     <gender> </gender> <profession> </profession> <head> </head>
     <body> </body> <legs> </legs> <voice> </voice>
     <home-town> </home-town> ... </person>
Mary female
Peter male
Jack male
<person> <gender> male </gender> <chin> beard </chin>
     <voice> deep </voice> </person>
<person> <gender> female </gender> <chin> no-beard </chin>
     <voice> high </voice> </person>
<profession> doctor has-stethoscope ... </profession>
<profession> lawyer has-law-books ... </profession>
<profession> merchant has-warehouse ... </profession>
<profession> thief is-light-fingered ... </profession>
<head> <hair> </hair> <eyes> </eyes> <nose> </nose> <mouth>
     </mouth> <chin> </chin> ... </head>
<hair> <hair-colour> </hair-colour> <length> </length> </hair>
<eyes> blue </eyes>
<eyes> brown </eyes>
<eyes> green </eyes>
<hair-colour> red </hair-colour>
<hair-colour> black </hair-colour>
<hair-colour> fair </hair-colour>
<jack1> <person> <first-name> Jack </first-name>
     <last-name> Jones </last-name>
     <home-town> Dorking </home-town> </person> </jack1>
```

Fig. 3. A set of SP patterns providing a partial description of the class 'person' and associated concepts. Patterns that are too long to fit on one line are indented on the second and subsequent lines

there would be similar patterns describing the structure of 'body' and 'legs'. In a similar way, each component of 'head', 'body' and 'legs' may itself be broken down into parts and subparts, thus yielding a complete part-whole hierarchy for the class 'person'.

The next six patterns in Fig. 3 give a set of alternative descriptions for 'eyes' and 'hair-colour', and the last pattern describes a specific person, Jack Jones.

3.1 The Best Multiple Alignment and Its Interpretation

Figure 4 shows the best alignment found by SP62 with the patterns from Fig. 3 in Old and the set of patterns {'has-stethoscope', '<hair-colour> fair </hair-colour>', <first-name> Jack </first-name>', '<home-town> Dorking </home-town>', 'has-black-bag', '<eyes> blue </eyes>'} in New. In this example, it has been necessary to use abbreviations for symbols to allow the alignment to fit into the printed page. The key to these abbreviations is shown in the caption to the figure. The fact that alignments can often grow to be quite large underscores the point that the building of multiple alignments is primarily a computational technique and users of the system would not normally see results presented in this form.

As before, this alignment may be interpreted as the result of a process of recognising some unknown entity with the attributes described in New. And, as before, it may also be interpreted in terms of information retrieval and probabilistic inference.

In this example, the unknown entity is recognised as being 'Jack Jones', a member of the class 'person'. At the same time, the entity is recognised as belonging to the subclasses 'male' (gender) and 'doctor' (profession). As with our previous example, our unknown entity inherits all the attributes of the classes to which it has been assigned. Notice that this inheritance works even though 'male' is not a subclass of 'doctor' or *vice versa*. In the jargon of object-oriented design, this is an example of 'multiple inheritance' or cross-classification.

Notice how John's male gender has been inferred from his name (via the association shown in column 8) and how the inference that he is male leads to the further inferences that he has a deep voice and could grow a beard (column 6).

Notice also how the alignment provides for inter-connections between different levels in the class hierarchy. The basic structure of a person is described in columns 2, 3 and 4 but details of that structure such as 'beard' and 'deep' voice are provided by the pattern for male gender shown in column 6.

3.2 Discussion

The two examples that have been presented so far capture the essentials of the way in which the SP system may be used for the representation and processing of class hierarchies, part-whole hierarchies and their integration.

Some readers may feel uneasy that the ideas have been presented without using familiar kinds of mathematical or logical notation. There may be an expectation that our informal concepts such as 'class', 'subclass', 'part', 'whole' and so on should be defined using traditional kinds of formal notation and that there should be formal proofs that these concepts are indeed captured by the SP system.

The alternative and more direct approach favoured here recognises that SP patterns and symbols are themselves a formal notation and we need only show how our informal concepts may be represented using this notation. In the remainder of this subsection, we briefly review a range of familiar concepts and the ways in which they may be expressed in the SP system.

Literal

In our first example, symbols like 'eats', 'breathes', 'has-senses', 'furry' and 'warm-blooded' may each be regarded as a 'literal' description of an attribute of some entity or class.

Variable, Value and Type

A pair of neighbouring symbols like '<animal> </animal>' in our first example or '<hair-colour> </hair-colour>' in our second example may be regarded as a 'variable' that may receive a 'value' by alignment of patterns, as illustrated here:

```
0       1        2        3       4       5       6        7        8    9
                                                           <j1>
                 <pn> --------------------------- <pn> --- <pn>
<fn> ----------- <fn> ------------------------------------ <fn>
Jk ------------------------------------------------------- Jk ----- Jk
</fn> ---------- </fn> ----------------------------------- </fn>
                 <ltn> ----------------------------------- <ltn>
                                                           Jn
                 </ltn> ---------------------------------- </ltn>
                 <gnd> -------------------------- <gnd>
                                                  ml -------------- ml
                 </gnd> ------------------------- </gnd>
        <pfn> -- <pfn>
        dc
hs ---- hs
        ...
        </pfn> - </pfn>
                 <hd> --- <hd>
                          <hr> -- <hr>
<hc> ---------------------------- <hc> -- <hc>
fr -------------------------------------- fr
</hc> --------------------------- </hc> - </hc>
                                  <ln>
                                  </ln>
                          </hr> - </hr>
<es> -------------------- <es> ---------------------------------------- <es>
bl -------------------------------------------------------------------- bl
</es> ------------------- </es> --------------------------------------- </es>
                          <ns>
                          </ns>
                          <mt>
                          </mt>
                          <cn> ------------------ <cn>
                                                  bd
                          </cn> ----------------- </cn>
                          ...
                 </hd> -- </hd>
                 <bdy>
                 </bdy>
                 <lg>
                 </lg>
                 <vc> -------------------------- <vc>
                                                  dp
                 </vc> ------------------------- </vc>
<ht> ----------- <ht> ------------------------------------ <ht>
Dg ------------------------------------------------------- Dg
</ht> ---------- </ht> ----------------------------------- </ht>
hb
                 ...
                 </pn> ------------------------- </pn> -- </pn>
                                                           </j1>

0       1        2        3       4       5       6        7        8    9
```

Fig. 4. The best alignment found by SP62 with the patterns from Fig. 3 in Old and New patterns as described in the text. *Key*: bd = beard, bdy = body, bl = blue, cn = chin, dc = doctor, Dg = Dorking, dp = deep, es = eyes, fn = first-name, fr = fair, gnd = gender, hb = has-black-bag, hc = hair-colour, hd = head, hr = hair, hs = has-stethoscope, ht = home-town, j1 = jack1, Jk = Jack, Jn = Jones, lg = legs, ln = length, ltn = last-name, ml = male, mt = mouth, ns = nose, pfn = profession, pn = person, vc = voice

```
0 ... <X>      </X> ... 0
       |         |
1     <X> V1 </X>       1.
```

In this example, ‘V1’ is a value assigned, by alignment, to the variable ‘<X> </X>’. The ‘type’ of a variable—the set of alternative values that it may take—may be expressed in a set of patterns such as ‘<X> V1 </X>’, ‘<X> V2 </X>’ and ‘<X> V3 </X>’ or in patterns like ‘<hair-colour> red </hair-colour>’, ‘<hair-colour> black </hair-colour>’ and ‘<hair-colour> fair </hair-colour>’, shown in Fig. 3.

Reference or Pointer

A widely-used device in computing systems is a ‘reference’ or ‘pointer’ from one structure to another. In the SP system, a comparable effect may be achieved by the alignment of symbols between patterns. For example, the pair of symbols ‘<mammal> </mammal>’ in column 2 of Fig. 2 may be seen as a reference to the pattern ‘<mammal> <animal> </animal> furry warm-blooded ... </mammal>’ in column 3.

Class, Subclass and Instance

In the SP system, any ‘class’, ‘subclass’ or ‘instance’ (‘object’) may be represented with a pattern as illustrated in our two examples. The relationship between a class and one of its subclasses or between a class or subclass and one of its instances is defined by the matching of symbols between patterns, exactly as described for ‘references’ and ‘variables’.

Since there is no formal distinction between ‘class’ (or ‘subclass’) and ‘instance’ this means that a pattern that represents an instance may also serve as a class wherever that may be appropriate or necessary. At first sight, this seems to conflict with the idea that an instance is a singleton (like ‘Tibs’ in column 1 of Fig. 2) whereas a class (like ‘cat’ or ‘mammal’) represents a set of instances. But we should remember that our concept of a cat such as ‘Tibs’ is a complex thing that derives from many individual ‘percepts’ such as “Tibs stalking a mouse”, “Tibs sleeping”, “Tibs cleaning himself”, and so on. Within the SP system, we can capture these manifestations of Tibs’s existence and their relationship to the overarching concept of ‘Tibs’ in precisely the same way as we can describe the relationship between any other class and one of its subclasses or instances, as described above.

In systems that make a formal distinction between instances (or ‘objects’) and classes, there is a need—not always satisfied in any given system—for a concept of ‘metaclass’:

> “If each object is an instance of a class, and a class is an object, the [object-oriented] model should provide the notion of *metaclass*. A metaclass is the class of a class.” [3, p. 43].

By the same logic, we should also provide for 'metametaclasses', 'metametameta-classes', and so on without limit. Because the SP system makes no distinction between 'object' and 'class', there is no need for the concept of 'metaclass' or anything beyond it. All these constructs are represented by patterns.

Parts, Wholes and Attributes

In the Simula computer language and most object-oriented systems that have come after, there is a distinction between 'attributes' of objects and 'parts' of objects. The former are defined at compile time while the aggregation of parts to form wholes is a run-time process. This means that the inheritance mechanism—which operates on class hierarchies that are defined at compile time—applies to attributes but not to parts.

In the SP system, the distinction between 'attributes' and 'parts' disappears. Parts of objects can be defined at any level in a class hierarchy and inherited by all the lower level. There is seamless integration of class hierarchies with part-whole hierarchies.

Intensional and Extensional Representations of Classes

There is a long-standing tradition that a concept, class or category may be described 'intensionally'—in terms of its attributes—or 'extensionally' by listing the things that belong in the class.

The SP system allows both styles of description and they may be freely intermixed. A class like the category 'person' may be described intensionally with patterns that describe the attributes of a typical person (like the three patterns that begin with '<person>' in Fig. 3) or it may be described extensionally with a set of patterns like this:

```
<person> Mahatma Gandhi </person>
<person> Abraham Lincoln </person>
<person> John Lennon </person>
...
```

Other examples of extensional categories are the alternative values for eye colour shown in Fig. 3 and kinds of profession shown in the same figure.

Polythetic or 'Family Resemblance' Classes

A characteristic feature of the 'natural' categories that we use in everyday thinking, speaking or writing is that they are 'polythetic' [16] or 'family resemblance' concepts. This means that no single attribute of the class need necessarily be found in every member of the class and none of the attributes are necessarily exclusive to the class. Since this type of concept is a prominent part of our thinking, knowledge-based systems should be able to accommodate them.

The SP system provides two main mechanisms for representing and processing polythetic classes:

- At the heart of the system for building multiple alignments is an improved version of 'dynamic programming' [15] that allows the system to find good partial matches between patterns as well as exact matches (the algorithm is described in [20]). This means that an unknown entity may be recognised as belonging to a given category when only a subset of the attributes of the category have been matched to features of the unknown entity. And this recognition may be achieved even though some of the features of the given category are also found in other categories.
- The system can be used to model context-free and context-sensitive rules (see [23]) and such rules may be used to define polythetic categories. For example, if 'A', 'B', 'C' and 'D' are 'attributes' found in one or more categories, a polythetic category 'X' may be defined by the following re-write rules:

```
X -> 1 2
1 -> A
1 -> B
2 -> C
2 -> D
```

  These rules define the set of strings {'AC', 'AD', 'BC', 'BD'}, a class in which no single attribute is found in every member of the class. If any one of those attributes is found in any other class, then, in terms of the definition, X is polythetic.

4 Other Aspects of Intelligence

So far, the focus has been mainly on the application of the SP system to the representation and processing of ontological knowledge. This section briefly describes other capabilities of the SP system for intelligent representation and processing of knowledge. More detail may be found in [27] and earlier publications cited there.

4.1 Representation of Knowledge

The examples we have seen so far illustrate some of the expressive power of SP patterns within the SP framework. A variety of other systems can be modelled including context-free grammars, context-sensitive grammars, networks, trees, if-then rules, and tables. Some examples will be seen below and others may be found in [22, 23, 25, 27]. The simple, uniform nature of SP patterns means that these different styles of knowledge representation can be freely intermixed in a seamless manner.

In the SP framework, there is no formal distinction between syntax and semantics. Semantic constructs may be modelled within the system, each one associated with its corresponding syntax.

4.2 Fuzzy Pattern Recognition and Best-Match Information Retrieval

As previously noted, SP62 incorporates an improved version of 'dynamic programming' and this gives it a robust ability to recognise patterns or retrieve stored information despite errors of omission, commission or substitution. Although this capability is needed to build alignments like those shown in Figs. 2 and 4, this aspect of the model is illustrated more clearly in other examples that may be found in [22, 24, 27].

4.3 Probabilistic Reasoning

One of the main strengths of the SP62 model is its capability for probabilistic reasoning, including probabilistic 'deduction', chains of reasoning, abduction, nonmonotonic reasoning, and 'explaining away'.[2] These applications of the model and the method of calculating probabilities are described quite fully in [22]. To give the flavour of these applications, this section describes one simple example of the way in which the model can support nonmonotonic reasoning.

If we know that Tweety is a bird, we may infer that Tweety can probably fly but that there is a possibility that Tweety might be a penguin or some other kind of flightless bird. If we are then told that Tweety is a penguin, we will revise our ideas and conclude with some confidence that Tweety cannot fly. This kind of reasoning is 'nonmonotonic' because later information can modify earlier conclusions (for a useful account of this topic, see [7]). By contrast, systems for 'classical' logic are designed to be 'monotonic' so that conclusions reached at one stage cannot be modified by any information arriving later.

Figure 5 shows the two best alignments found by SP62 with a set of patterns describing kinds of birds in Old and, in New, the pattern 'bird Tweety' (which may be interpreted as a statement that Tweety is a bird). The first alignment (a) confirms that Tweety is a bird and suggests, in effect, that Tweety can probably fly—column 3 in this alignment expresses this default assumption about birds. The second alignment (b) also confirms that Tweety is a bird but suggests, in effect, that Tweety might be a penguin and, as such, he would not be able to fly.

Each pattern in Old has an associated frequency of occurrence in some domain and, using these figures, SP62 is able to calculate relative probabilities for alignments. The default frequency value for any pattern is 1 but, in this example, frequency values were assigned to the patterns as very approximate estimates of the frequencies with which one might encounter birds, penguins and so on in the real world. For this toy example (which does not recognise ostriches, kiwis etc), SP62 calculates the probability that Tweety can fly as 0.84 and the probability that Tweety is a penguin and that he cannot fly as 0.16.

What happens if, instead of knowing merely that Tweety is a bird, we are given the more specific information that Tweety is a penguin? With SP62, we can model this kind of situation by running the model again and replacing 'bird Tweety' in New with the pattern 'penguin Tweety'—which we can interpret as a statement that

[2] For an explanation of this last idea, see [13].

```
0          1          2         3

                                Default
                      Bd ----   Bd
bird ---------------  bird
           name ---   name
Tweety - Tweety
           #name --   #name
                      f -----   f
                                can-fly
                      #f ----   #f
                      ...
                      #Bd ---   #Bd
                                #Default

0          1          2         3

(a)

0          1          2         3

                                P
                                penguin
                      Bd ----   Bd
bird ---------------  bird
           name ---   name
Tweety - Tweety
           #name --   #name
                      f -----   f
                                cannot-fly
                      #f ----   #f
                      ...
                      #Bd ---   #Bd
                                ...
                                #P

(b)

0          1          2         3
```

Fig. 5. The two best alignments found by SP62 with patterns describing kinds of birds in Old and the pattern 'bird Tweety' in New

Tweety is a penguin. In this case, there is only one alignment that matches both symbols in the pattern in New. From this alignment—shown in Fig. 6—we may conclude that Tweety cannot fly. The probability in this case is 1.0 because there is no other alignment that accounts for all the information in New.

4.4 Natural Language Processing

Much of the inspiration for the SP framework has been a consideration of the structure of natural languages and how they may be learned [19]. Here, a slightly quirky example is presented to show how the parsing of natural language may be modelled

```
0         1        2       3

                           P
penguin ----------------- penguin
                   Bd ---- Bd
                   bird
          name --- name
Tweety -- Tweety
          #name -- #name
                   f ----- f
                           cannot-fly
                   #f ---- #f
                   ...
                   #Bd --- #Bd
                           ...
                           #P

0         1        2       3
```

Fig. 6. The best alignment found by SP62 with the same patterns in Old as were used for Fig. 5, and the pattern 'penguin Tweety' in New

within the SP framework. SP62 was run with patterns representing grammatical rules in Old and, in New, the second sentence from Groucho Marks's "Time flies like an arrow. Fruit flies like a banana." The syntactic ambiguity of that sentence can be seen in the two best alignments found by the program, shown in Fig. 7. The horizontal format for the alignments is more appropriate here than the vertical format used in previous alignments and, for this kind of application, the use of angle brackets as boundary markers (Sect. 2.2) is slightly neater than the '<tag> ... </tag>' style.

Of course, 'Fruit flies like a banana' is ambiguous at the semantic level as well as in its syntax, and the kind of syntactic analysis shown in Fig. 7 falls short of what is needed for the understanding of natural languages. However, the SP framework has been developed with the intention that it should allow the representation of non-syntactic 'semantic' knowledge and its integration with syntax. Preliminary work in this area—not yet published—confirms this expectation.

The alignments shown in Fig. 7 may suggest that the system has only the expressive power of a context-free phrase-structure grammar (CF-PSG), not sufficient in itself to handle the syntactic subtleties of natural languages. However, other examples may be found in [23] showing how the system can handle 'context sensitive' features such as number agreements and gender agreements in French and the interesting pattern of inter-locking constraints found in English auxiliary verbs.

What about the production of natural language? A neat feature of the SP framework is that it supports the production of language as well as its analysis. Given a compressed, encoded representation of a sentence, the system can recreate the sentence from which it was derived (see [23] and also [27]). As noted above, preliminary work has shown that, within the SP framework, it is possible to integrate syntax with semantics. Using these integrated sets of SP patterns, it is possible to generate sentences from meanings.

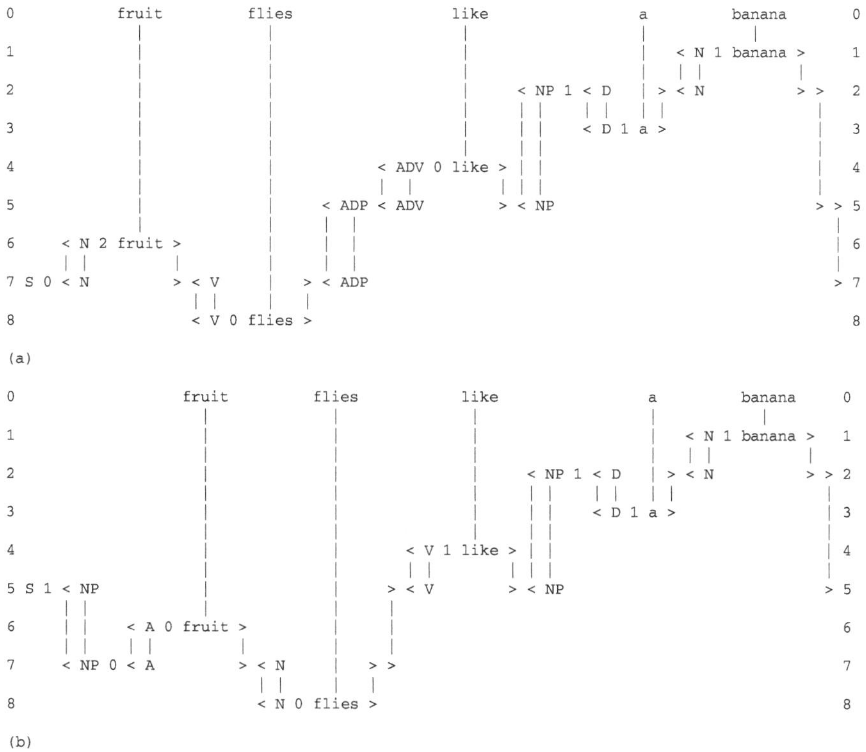

Fig. 7. The two best alignments found by SP62 with patterns in Old representing grammatical rules and the ambiguous sentence 'fruit flies like a banana' in New

4.5 Planning and Problem Solving

As it stands now, the SP62 model can be applied to simple kinds of planning and simple kinds of problem solving. Figure 8 shows the best alignment found by SP62 with the pattern 'Beijing Edinburgh' in New and a set of patterns in Old, each one of which represents a flight between two cities. The pattern in New may be interpreted as a request to find a route between Beijing and Edinburgh and the alignment represents one possible answer. If alternative routes are possible, the model normally finds them.

To be realistic, the system would have to be able to handle numeric information such as costs, distances and times. In principle this can be done within the SP framework [22] but the details have not yet been worked out.

The SP62 model may also be applied to the solution of geometric analogy problems of the form 'A is to B as C is to ?'. In order to use the model, each geometric pattern must be translated into a textual pattern such 'small circle inside large triangle'. Given a problem expressed in this form, SP62 solves it quite easily [22].

Further work is needed to explore the strengths and limitations of the SP framework in applications like these.

```
0              1           2           3             4

Beijing --- Beijing
               Delhi --- Delhi
                           Zurich ------------ Zurich
                                       London ---- London
Edinburgh -------------------- Edinburgh

0              1           2           3             4
```

Fig. 8. The best alignment found by SP62 with patterns in Old representing air links between cities and the pattern 'Beijing Edinburgh' in New

4.6 Unsupervised Learning

As we saw in Sect. 2, the SP framework is designed to receive New information and add it, in compressed form, to its repository of Old information. Thus, in its overall organisation, the system is designed to learn from its environment.

This aspect of the framework is now realised in the SP70 model [26,28]. Given a set of simple sentences, SP70 can abstract a plausible grammar for those sentences. This is done without the need for error correction by a 'teacher', without the need for 'negative' samples (sentences that are marked as 'wrong') and without the need for examples to be presented in any particular order (*cf.* [8]). In short, SP70 is a system for *unsupervised* learning.

Although the SP framework appears to provide a sound basis for unsupervised learning, more work is needed to resolve some residual problems and realise the full potential of the framework in a working system.

Although the SP70 model has been developed using examples with a 'linguistic' flavour, similar principles apply to other kinds of knowledge. When the system is more fully developed, potential applications include:

- The abstraction of general rules from a body of knowledge for subsequent use in probabilistic reasoning.
- The conversion of 'raw' data (or other badly-organised body of knowledge) into a 'well structured' database that preserves all the non-redundant information in the original.
- The integration of two or more bodies of knowledge that are similar but not the same.

5 Comparison with Alternative Approaches

One of the most popular approaches to the representation and processing of ontological knowledge is a family of languages developed for the Semantic Web, including the Resource Description Framework (RDF), RDF Schema (RDFS), the Ontology Inference Layer (OIL) and DAML+OIL (see, for example, [5, 6, 10, 12]). RDF provides a syntactic framework, RDFS adds modelling primitives such as 'instance-of' and 'subclass-of', and DAML+OIL provides a formal semantics and reasoning capabilities based on Description Logics (DLs), themselves based on predicate logic. For the sake of brevity, these languages will be referred to as 'RDF+'. In this section, RDF+ is compared with the SP framework, focussing mainly on the differences between them.

5.1 Representation of Knowledge

The SP theory incorporates a theory of knowledge in which syntax and semantics are integrated and many of the concepts associated with ontologies are implicit. There is no need for the explicit provision of constructs such as 'Class', 'subClassOf', 'Type', 'hasPart', 'Property', 'subPropertyOf', or 'hasValue'. Classes of entity and specific instances, their component parts and properties, and values for properties, can be represented in a very direct, transparent and intuitive way, with smooth integration of class-inclusion relations and part-whole relations, as described above. In short, the SP approach allows ontologies to be represented in a manner that appears to be simpler and more comprehensible than RDF+, both for people writing ontologies and for people reading them.

5.2 Recognition and Classification

RDF+ has been designed with the intention that authors of web pages would specify the ontological structure of each page alongside its content. However, with current versions of RDF+, this is far from easy and this may prove to be a severe barrier to the uptake of these technologies [9].

In this connection, the SP framework offers an interesting possibility—that the capabilities of the system for 'fuzzy' recognition of concepts at multiple levels of abstraction may be exploited to achieve automatic or semi-automatic assignment of web pages to pre-defined categories expressed in the form of SP patterns. Given a set of such patterns in its repository of 'Old' information, the system may find one or more categorizations of a given web page if that page is presented to the system as New information. If this kind of automatic categorization of web pages can be realised, this should streamline the assignment of ontologies to web pages and thus smooth the path for the development of the Semantic Web.

5.3 Reasoning

Reasoning in RDF+ is based on predicate logic and belongs in the classical tradition of monotonic deductive reasoning in which propositions are either *true* or *false*. By contrast, the main strengths of the SP framework are in probabilistic 'deduction', abduction and nonmonotonic reasoning. It seems likely that both kinds of reasoning will be required in the Semantic Web, as discussed in the following subsections.

Probabilistic and Exact Reasoning

The dynamic programming at the heart of the SP system allows it to find good partial matches between patterns as well as exact matches. The building of multiple alignments involves a search for a global best match amongst patterns and accommodates errors of omission, commission and substitution in New information or Old information or both. This kind of capability—and the ability to calculate probabilities of inferences—is outside the scope of 'standard' reasoning systems currently developed for RDF+ (see, for example, [10]). It has more affinity with 'non-standard inferences' for DLs (see, for example, [1,4]).

Given the anarchical nature of the Web, there is a clear need for the kind of flexibility provided in the SP framework. It may be possible to construct ontologies in a relatively disciplined way but ordinary users of the web are much less controllable. The system needs to be able to respond to queries in the same flexible and 'forgiving' manner as current search engines. Although current search engines are flexible, their responses are not very 'intelligent'. The SP framework may help to plug this gap.

Deduction and Abduction

With the Semantic Web, we may wish to reason deductively in some situations (e.g., a flat battery in the car means that it will not start) but there will also be occasions when we wish to reason abductively (e.g., possible reasons for the car not starting are a flat battery, no fuel, dirty plugs etc). For each alternative identified by abductive reasoning, we need some kind of measure of probability.

Although classical systems can be made to work in an abductive style, an ability to calculate probabilities or comparable measures needs to be 'bolted on'. By contrast, the SP system accommodates 'backwards' styles of reasoning just as easily as 'forwards' kinds of reasoning and the calculation of probabilities derives naturally from the minimum length encoding principles on which the system is founded.

Monotonic and Nonmonotonic Reasoning

If we know that all birds can fly and that Tweety is a bird then, in classical logic, we can deduce that Tweety can fly. The 'monotonic' nature of classical logic means that this conclusion cannot be modified, even if we are informed that Tweety is a penguin.

In the Semantic Web, we need to be able to express the idea that *most* birds fly (with the default assumption that birds can fly) and we need to be able to deduce that, if Tweety is a bird, then it is *probable* that Tweety can fly. If we learn subsequently that Tweety is a penguin, we should be able to revise our initial conclusion.

This kind of nonmonotonic reasoning is outside the scope of classical logic but is handled quite naturally by the SP framework, as outlined in Sect. 4.3 (see also [22,24]).

6 Conclusion

One of the strengths of the SP framework is that it allows ontological knowledge to be represented in a manner that is relatively simple and comprehensible. It has an ability to cope with uncertainties in knowledge and it can perform the kinds of probabilistic reasoning that seem to be needed in the Semantic Web.

The SP framework also has strengths in other areas—such as natural language processing, planning and problem solving, and unsupervised learning—that may prove useful in the future development of the Semantic Web.

Acknowledgements

I am very grateful to Manos Batsis, Pat Hayes, Steve Robertshaw and Heiner Stuckenschmidt for constructive comments that I have received on these ideas.

References

1. F. Baader, R. Küsters, A. Borgida, and D. L. McGuinness. Matching in description logics. *Journal of Logic and Computation*, 9(3):411–447, 1999.
2. T. Berners-Lee, J. Hendler, and O. Lassila. The Semantic Web. *Scientific American*, May 2001.
3. E. Bertino, B. Catania, and G. P. Zarri. *Intelligent Database Systems*. Addison-Wesley, Harlow, 2001.
4. S. Brandt and A.-Y. Turhan. Using non-standard inferences in description logics: what does it buy me? In *Proceedings of KI-2001 Workshop on Applications of Description Logic (KIDLWS'01)*, 2001.
5. J. Broekstra, M. Klein, S. Decker, D. Fensel, F. van Harmelin, and I. Horrocks. Enabling knowledge representation on the web by extending RDF schema. *Computer Networks*, 39:609–634, 2002.
6. D. Fensel, I. Horrocks, F. van Harmelin, D. L. McGuinness, and P. F. Patel-Schneider. OIL: an ontology infrastructure for the Semantic Web. *IEEE Intelligent Systems*, 16(2):38–45, 2001.
7. M. L. Ginsberg. AI and nonmonotonic reasoning. In D. M. Gabbay, C. J. Hogger, and J. A. Robinson, editors, *Handbook of Logic in Artificial Intelligence and Logic Programming: Nonmonotonic Reasoning and Uncertain Reasoning*, volume 3, pages 1–33. Oxford University Press, Oxford, 1994.
8. M. Gold. Language identification in the limit. *Information and Control*, 10:447–474, 1967.
9. P. Hayes. Catching the dreams. Technical report, 2002. Copy: www.aifb.uni-karlsruhe.de/ sst/is/WebOntologyLanguage/hayes.htm.
10. I. Horrocks. Reasoning with expressive description logics: theory and practice. In A. Voronkov, editor, *Proceedings of the 18th International Conference on Automated Deduction (CADE-18)*, volume 2392 of *Lecture Notes in Artificial Intelligence*, pages 1–15. Springer-Verlag, 2002.

11. M. Li and P. Vitányi. *An Introduction to Kolmogorov Complexity and Its Applications*. Springer-Verlag, New York, 1997.
12. D. L. McGuinness, R. Fikes, J. Hendler, and L. A. Stein. DAML+OIL: an ontology language for the Semantic Web. *IEEE Intelligent Systems*, 17(5):72–80, 2002.
13. J. Pearl. *Probabilistic Reasoning in Intelligent Systems*. Morgan Kaufmann, San Francisco, revised second printing edition, 1997.
14. J. Rissanen. Modelling by the shortest data description. *Automatica-J, IFAC*, 14:465–471, 1978.
15. D. Sankoff and J. B. Kruskall. *Time Warps, String Edits, and Macromolecules: the Theory and Practice of Sequence Comparisons*. Addison-Wesley, Reading, MA, 1983.
16. R. R. Sokal and P. H. A. Sneath, editors. *Numerical Taxonomy: the Principles and Practice of Numerical Classification*. W. H. Freeman, San Francisco, 1973.
17. R. J. Solomonoff. A formal theory of inductive inference. parts I and II. *Information and Control*, 7:1–22 and 224–254, 1964.
18. C. S. Wallace and D. M. Boulton. An information measure for classification. *Computer Journal*, 11(2):185–195, 1968.
19. J. G. Wolff. Learning syntax and meanings through optimization and distributional analysis. In Y. Levy, I. M. Schlesinger, and M. D. S. Braine, editors, *Categories and Processes in Language Acquisition*, pages 179–215. Lawrence Erlbaum, Hillsdale, NJ, 1988. Copy: www.cognitionresearch.org.uk/lang_learn.html#wolff_1988.
20. J. G. Wolff. A scaleable technique for best-match retrieval of sequential information using metrics-guided search. *Journal of Information Science*, 20(1):16–28, 1994. Copy: www.cognitionresearch.org.uk/papers/ir/ir.htm.
21. J. G. Wolff. 'Computing' as information compression by multiple alignment, unification and search. *Journal of Universal Computer Science*, 5(11):777–815, 1999. Copy: http://arxiv.org/abs/cs.AI/0307013.
22. J. G. Wolff. Probabilistic reasoning as information compression by multiple alignment, unification and search: an introduction and overview. *Journal of Universal Computer Science*, 5(7):418–462, 1999. Copy: http://arxiv.org/abs/cs.AI/0307010.
23. J. G. Wolff. Syntax, parsing and production of natural language in a framework of information compression by multiple alignment, unification and search. *Journal of Universal Computer Science*, 6(8):781–829, 2000. Copy: http://arxiv.org/abs/cs.AI/0307014.
24. J. G. Wolff. Information compression by multiple alignment, unification and search as a framework for human-like reasoning. *Logic Journal of the IGPL*, 9(1):205–222, 2001. First published in the *Proceedings of the International Conference on Formal and Applied Practical Reasoning (FAPR 2000)*, September 2000, ISSN 1469–4166. Copy: www.cognitionresearch.org.uk/papers/pr/pr.htm.
25. J. G. Wolff. Mathematics and logic as information compression by multiple alignment, unification and search. Technical report, CognitionResearch.org.uk, 2002. Copy: http://arxiv.org/abs/math.GM/0308153.
26. J. G. Wolff. Unsupervised learning in a framework of information compression by multiple alignment, unification and search. Technical report, CognitionResearch.org.uk, 2002. Copy: http://arxiv.org/abs/cs.AI/0302015.
27. J. G. Wolff. Information compression by multiple alignment, unification and search as a unifying principle in computing and cognition. *Artificial Intelligence Review*, 19(3):193–230, 2003. Copy: http://arxiv.org/abs/cs.AI/0307025.

28. J. G. Wolff. Unsupervised grammar induction in a framework of information compression by multiple alignment, unification and search. In C. de la Higuera, P. Adriaans, M. van Zaanen, and J. Oncina, editors, *Proceedings of the Workshop and Tutorial on Learning Context-Free Grammars*, pages 113–124, 2003. This workshop was held in association with the 14th European Conference on Machine Learning and the 7th European Conference on Principles and Practice of Knowledge Discovery in Databases (ECML/PKDD 2003), September 2003, Cavtat-Dubrovnik, Croata. Copy: http://arxiv.org/abs/cs.AI/0311045.

Part II

Fuzzy Logic in Ontologies and Semantic Web

Dynamic Services for Open Ambient Intelligence Systems

Giovanni Acampora and Vincenzo Loia

Università degli Studi di Salerno
Dipartimento di Matematica e Informatica,
via Ponte don Melillo - 84084 Fisciano (Salerno), Italy
{gacampora, loia}@unisa.it

1. Introduction

Ambient intelligence (AmI) [1] provides a wide-ranging vision on how the Information Society will evolve, since the goal is to conceive platforms for seamless delivery of services and applications making them effectively invisible to the user. This is possible by gathering best practices from Ubiquitous Computing, Ubiquitous Communication, and Intelligent User Friendly Interfaces areas. This convergence will lead to smart environments that surround people in pro-active way: the environment reacts autonomously to people by using intelligent intuitive interfaces (often invisible) that are embedded in all kinds of chip-enriched objects (furniture, clothes, vehicles, roads and other smart materials). This means that computing and networking technology is everywhere, embedded in everyday objects in order to automate tasks or enhance the environment for its occupants.

This objective is achievable if the environment is capable to learn, build and manipulate user profiles considering from a side the need to clearly identify the human attitude (AmI is known to work for people not for users) and from the other the ubiquity of the possible services. Of course, this involves several research trends:

- new generation of devices, sensors, and interfaces [2];
- new generation of processors and communication infrastructures [3];

G. Acampora and V. Loia: *Dynamic Services for Open Ambient Intelligence Systems*, StudFuzz **204**, 105–122 (2006)
www.springerlink.com

- new "intelligence" empowerment, provided by software agents embedded in the devices [4].

Sensors are connected in a wired (or unwired) way for a certain application, and in general the association sensor-activity is too rigid to enable a flexible or dynamic utilization. The deep exigency of abstraction and uniform interface access stimulates the research in providing abstract and problem-independent description model, useful for interoperability and adaptive control strategies. This issue, crucial for intelligent environment applications, is not fully supported by actual available technologies: JINI [5] and UPnP [6] are two popular examples of network specifications for easy connection of home information appliances and computers, but they are rather primitive and demand a lot of complex work (in terms of several software layers) in order to bridge the framework side with the sensor software level.

This paper is devoted to discuss the problems related to the design of AmI environments and to demonstrate how is possible to obtain autonomy, independence and distribution of the computational resources by means of hybrid approach. The autonomy is realized using the well-know fuzzy logic theory which permits to control, in automatic way, the different devices composing AmI framework; the XML technologies allow to realize the independence feature; the web services technologies allow to obtain the distribution properties.

2. The Starting Point

The mail goal in the design of AmI environment is to automate some of the tasks currently performed by humans possibly with improvements in efficiency or quality of service. Of course, an ample variety of approaches may concur to achieve the aforementioned objective. Many works face the problem to collect and integrate information about the activities that occur within the environment [7-9]: the detection accuracy is a critical trade-off between quantity and quality of recognized activities. Other efforts are concentrating in identifying and tracking humans as they move about the environment [10-12]. Most of the works reported in literature are based on the estimation off human body postures in the context of video motion capture. Many and complex problems arise, most of these of graphical nature. It's very hard to define a unique "abstract" model for human tracking, for instance human walking and visual expressivity demand two different approaches. Our starting point investigates on a general approach for a high-level management of sensors and actuator for Ambient Intelligence Environment. Control strategy plays a fundamental role: our choice is in using fuzzy technology, not only to realize the device functionality, but also to define a unique framework by gaining benefits in terms of generalization, complexity reduction, and flexibility. Generally, an AmI system is characterized by four key facets:

• Embedding. Devices are (wired or unwired) plugged into the network. Some of the devices are "simple" sensors, other ones are "actuator" owning a crunch of control activity on the environment. The strong heterogeneity makes difficult a uniformed policy-based management.

• Context awareness. Roughly, the system should own a certain ability to recognize people and the situational context.

•Personalization. AmI environments are designed for people, not generic users. This means that the system should be so flexible to tailor itself to meet human needs.

•Adaptivity. The system, being sensible to the user's feedback, is capable to modify the corresponding actions have been or will be performed.

In all of the previous issues, the functional and spatial distribution of components and tasks is a natural thrust to employ the agent paradigm to design and implement AmI environments. AmI are logically decentralized subsystems that can be viewed as adaptive agents involved in parallel (semi)local interactions. These interactions give rise to: user's actions detection by means of the network of sensor-agents, control activities in the environment by means of the pool of distributed actuator-agents, behavioral interactions with the people by means of user-interface agents.

Even though we report an increasing tendency in adopting the agent paradigm in AmI applications [13,14,15], a challenge remains still to solve: how the agent paradigm can support a flexible and dynamic reconfiguration of control services by balancing the hardware constraints with an abstract and uniform approach.

The rest of the paper is devoted to discuss this problem and we demonstrate how is possible to reach this equilibrium by a synergistic mixing of three approaches: markup languages, fuzzy control strategies and mobile agents.

3. Fuzzy Markup Language

Extensible Markup Language (XML) [16] is a simple, very flexible text format derived from SGML (ISO 8879). Originally designed to meet the challenges of large-scale electronic publishing, nowadays XML plays a fundamental role in the exchange of a wide variety of data on the Web, allowing designers to create their own customized tags, enabling the definition, transmission, validation, and interpretation of data between applications, devices and organizations. If we use XML, we take control and responsibility for our information, instead of abdicating such control to product vendors. This is the motivation under Fuzzy Markup Language (or FML) [17][18] proposal. The real implementation of fuzzy controllers is realized using XSLT tools which permit to translate the XML-based code into computer programs coded in a specific language The proposed solution translates the FML programs into Java programs embedded in web services components in order to realize the distribution feature and to strong the independence concept.

Figure 1 shows the architecture framework structured by different layers and how these layers realize the different features.

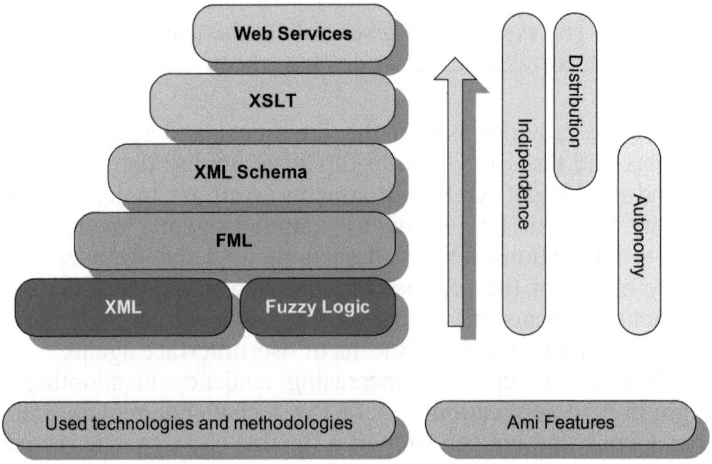

Figure 1 - AmI Architecture

In following sections we discuss more in detail on the layers composing the Ambient Intelligence framework.

FML language is a novel computer language used to model control systems based on fuzzy logic theories. Figure 2 illustrates the role played by FML in an AmI environment. The main feature of FML is the transparency property: the FML programs can be executed on different hardware without additional efforts. This property is fundamental in ubiquitous computing environment where computers are available throughout the physical environment and appear invisible and transparent to the user. Similarly to the other modern computer languages it is possible to define the FML language by describing the set of symbols (lexemes of language) used to write its programs and what its programs look like (the syntax of the language).

The **<FuzzyControl>** is the root tag of each FML program, that is, the opening tag of each FML program. **<FuzzyControl>** uses three tag types: *defuzzifyMethod, ip* and *name*. defuzzifyMethod defines the defuzzification method used in modeled controller; ip is used to define the location of controller in the computer network; name is used to identify, univocally, the controller . It is necessary to define the set of lexical elements useful to represent the main components of fuzzy controllers: knowledge base and rule base. The fuzzy knowledge base is defined by means of the tag **<KnowledgeBase>** which encloses the set of fuzzy concepts used to model the fuzzy rule base. This tag uses a set of nested tags: **<FuzzyVariable>**, **<FuzzyTerm>** and a set of tags defining a shape of fuzzy sets, as explained below. **<FuzzyVariable>** defines the fuzzy concept, for example, "temperature"; **<FuzzyTerm>** defines a linguistic term describing the fuzzy concept, for example, "low temperature"; the set of tags defining the shapes of fuzzy sets are related to fuzzy terms. The attributes of **<FuzzyVariable>** tags are: *name, scale, domainLeft, domainRight, type, ip*. name defines the name of fuzzy concept, for instance, temperature; scale is used to define the scale used to measure the fuzzy concept, for instance, Celsius degree; domainLeft and domainRight are used to model the universe of discourse of fuzzy concept, that is, the set of real values related to fuzzy concept, for instance, [0°, 40°]; the position of fuzzy concept into rule (consequent part or antecedent part) is defined by type attribute; ip defines the position of fuzzy knowledge base in the computer network. **<FuzzyTerm>** allows to define the linguistic value associate with fuzzy concept. Fuzzy shape tags, used to complete the definition of fuzzy concept, are: **<TriangularShape>, <LinearShape>, <TrapezoidShape>, <SShape>, <ZShape>, <PIShape>.**

<RuleBase> defines the fuzzy rule base: it encloses the set, eventually empty, of fuzzy rules composing the modelled controller. This tag uses two attributes: *inferenceEngine* and *ip*. The former is used to define inference operator , the latter defines the network location of the set of rules used in fuzzy controller. In order to define the single rule the tag **<Rule>** is used. The tags used by **<Rule>** are: *id, connector, weight, ip*. The id attribute permits to identify the rule; connector is used to define the logical operator used to connect the different clauses in antecedent part; weight defines the importance of rule during inference engine time; ip defines the location of rule in the computer network. The definition of antecedent and consequent rule part is obtained used **<Antecedent>** and **<Consequent>** tags.

<ClauseA> and **<ClauseC>** tags are used to model the fuzzy clauses in antecedent and consequent part, respectively. In order to treat the operator "not" in fuzzy clauses **<ClauseA>** and **<ClauseC>** uses the boolean attribute "not". To complete the definition of fuzzy clause the **<Variable>, <Term>, <TSKParam>** have to be used. In particular the pair **<Variable>, <Term>** is used to define fuzzy clauses in antecedent and consequent part of Mamdani controllers rules and in antecedent part of TSK controllers rules, while, the pair **<Variable>, <TSKParam>** is used to model the consequent part of TSK controllers rules.

FML uses XML Schema to define the syntax of markup language. The purpose of an XML Schema is to define the legal building blocks of an XML document, just like a DTD, using a more complex syntax than DTD but obtaining more advantages, such as:

- XML Schemas are extensible to future additions;
- XML Schemas are richer and more useful than DTDs;
- XML Schemas are written in XML;
- XML Schemas support data types;
- XML Schemas support namespaces.

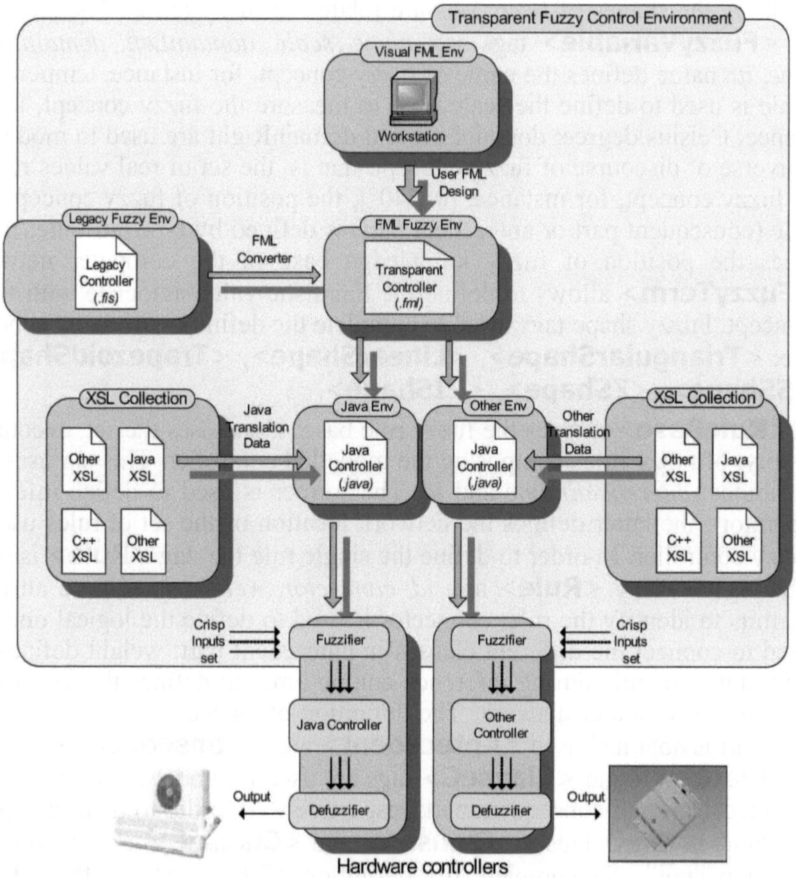

Figure 2 - Role of FML in AmI Architecture

The following XML code shows a portion of XML Schema defining the FML syntax:

```
<xs:sequence>
<xs:element name = "KnowledgeBase"
      type = "KnowledgeBaseType">
      <xs:annotation>
            <xs:documentation>
                  Fuzzy Concepts Collection
            </xs:documentation>
      </xs:annotation>
</xs:element>
<xs:element name="RuleBase"
      type="RuleBaseType">
      <xs:annotation>
        <xs:documentation>
            Fuzzy Rules Collection </xs:documentation>
      </xs:annotation>
</xs:element>
</xs:sequence>
<xs:attribute name="ControllerType">
      <xs:simpleType>
            <xs:restriction base="xs:string">
            <xs:pattern value="Mamdani|TSK"/>
            </xs:restriction>
      </xs:simpleType>
</xs:attribute>
<xs:attribute name="DefuzzificationMethod">
<xs:simpleType>
      <xs:restriction base="xs:string">
      <xs:pattern
      value="MinMaxMinMamdani|LarsonProduct"/>
      </xs:restriction>
</xs:simpleType>
</xs:attribute>
<xs:attribute name="IP">
<xs:simpleType>
      <xs:restriction base="xs:string">
      <xs:pattern
      value = "( (1?[0-9]?[0-9]|2[0-4][0-9]|25[0-5]).){3}(1?[0-
      9]?[0-9]|2[0-4][0-9]|25[0-5])"/>
      </xs:restriction>
</xs:simpleType>
</xs:attribute>
</xs:complexType>
```

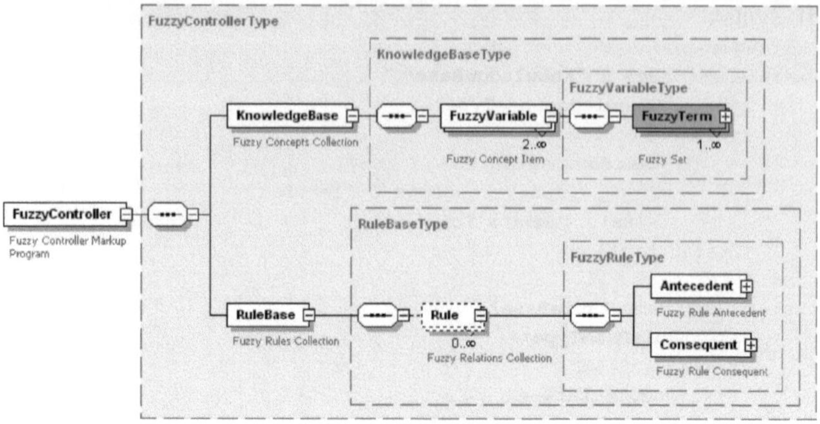

Figure 3 - A FOM portion

Using XML Schema rather than DTD allows to cover different problems, typical of programming languages, previously neglected in the first proposal of FML project: the semantic and the dynamicity. The semantic concepts in FML definition allows to define advanced computer languages concepts as fuzzy data type, fuzzy objects, etc.. Dynamicity concepts are intended to overcome the problems related to structure of program written by using the initial version of FML. In fact, old FML programs models the static structure of fuzzy controllers without to allows the run-time updating of the control systems. The only allowed modifications on fuzzy controllers coded in FML are possible by using extern algorithms able to rewrite opportunely the FML programs as discussed in [17, 18]. The semantic features expresses by XML Schema allows to define object-oriented computer markup language. In fact, XML Schema includes a hierarchical type system, using inheritance, pointers to other objects, and notions that data carries with it both a name (the element or object type name) and a more fundamental type (from the hierarchy). Using this XML Schema property is it possible to define a hierarchical objects model representing the different components of a fuzzy controller: the Fuzzy Objects Model (FOM). FOM offers the possibility to approach to different controller components and modify it during execution time. The "on-the-fly" modification is allowable by introducing a new concept: the Fuzzy Scripts. The Fuzzy Scripts concept is strongly related to HTML+Javascript concept; in fact, the used operation principle of Fuzzy Script is the same of JavaScript, i.e. to create, dynamically, HTML programs. In FML case, the dynamic platform allows to create fuzzy concepts, as variables and rules, starting from external information.

In order to implement the Fuzzy Scripts, the markup language uses two new tags: **<FuzzyControlHeader>** and **<FuzzyScript>**. The former permits to define a script declaration sections in fuzzy programs; this section contains a set of fuzzy scripts reusable in the rest of program. The latter encapsulates the script code. From a web point of view, the FOM corresponds to a web browser Document Objects Model (DOM), that is, the objects collection composing a HTML document and its visualization environment (browser). Using this analogy it is possible to define the Fuzzy Script Language, corresponding to JavaScript, as a scripting language able to interacts with FOM in order to create or modify the FML programs. Using FML script is possible to accelerate the fuzzy program realization, using automatic approach to generate new fuzzy elements as concepts and rules. The scripts used in FML uses the C/Java syntax and uses the same flow control construct used from C/Java languages. Moreover, the Fuzzy Script language uses new lexemes rather then C/Java in order to deal the fuzzy operators in direct way. For instance, is it possible to apply the basic Zadeh-Type operations on fuzzy concepts as intersection, union, and complement.

FML script can be used to model new fuzzy concepts starting from existing fuzzy concepts or to generate new fuzzy rules starting from existing fuzzy rules or from other knowledge sources.

FML scripts are executed and translated in pure markup language in direct fashion (see Figure 4), in the same way as JavaScripts are translated in HTML. FML scripts are useful to approach at FOM in order to create a fuzzy element (variable or rule) from scratch or to create a new fuzzy object starting from previous defined FOM objects, as variables or rules. In particular, this approach allows to create new concepts in dynamic way starting from the static definition modeled using a classical FML scheme.

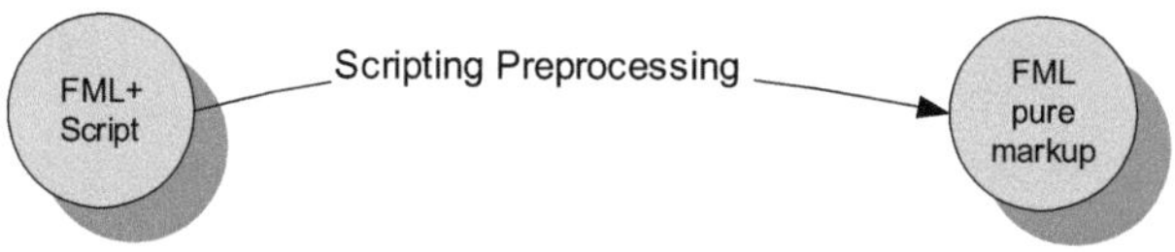

Figure 4 - FML script preprocessing

The following markup code represents an examples of oneShot script:

```
<FuzzyControl name = "Example">
<FuzzyHeader>
<FuzzyScript>
FuzzyScript calculateMediumTemperature() {
FuzzyVariable temperature =
  getFuzzyVariable("Temperature");
FuzzyTerm lowTemperature =
  temperature.getTerm("low");
```

```
FuzzyTerm highTemperature =
  temperature.getTerm("high");
FuzzyTerm mediumTemperature =
  new FuzzyTerm("medium");

mediumTemperature =
  ( not lowTemperature ) intersect ( not
         highTemperature);

temperature.addTerm(mediumTemperature);
} // endOfScript
</FuzzyScript>
</FuzzyHeader>

<KnowledgeBase>
<FuzzyVariable name = "Temperature"
       domainleft = "0" domainright = "100"
       scale = "Celsius">
<FuzzyTerm name = "low"> …</FuzzyTerm>
<FuzzyTerm name = "high"> …</FuzzyTerm>
<FuzzyScript> calculateMediumTemperature(); </FuzzyScript>
</FuzzyVariable>
</KnowledgeBase>
</FuzzyControl>
```

The pure FML code resulting from script preprocessing is:

```
<FuzzyControl name = "Example">
<KnowledgeBase>
       <FuzzyVariable name = "Temperature"
       domainleft = "0" domainright = "100"
       scale = "Celsius">
       <FuzzyTerm name = "low"> …</FuzzyTerm>
       <FuzzyTerm name = "high"> …</FuzzyTerm>
       <FuzzyTerm name = "medium">
              <UserShape>
                     <Point x = " " y = " "/>
                     <Point x = " " y = " "/>
                     <Point x = " " y = " " />
              </UserShape>
       </FuzzyTerm>
       </FuzzyVariable>
</KnowledgeBase>
</FuzzyControl>
```

4. FML Web Services

Web services, in the general meaning of the term, are services offered via the Web. In a typical Web services scenario, a business application sends a request to a service at a given URL using the SOAP [20] protocol over HTTP. The service receives the request, processes it, and returns a response. In our case, the services offered over HTTP protocol are control services. In particular, the fuzzy web services are computer programs, implementing a fuzzy controller, resident on the Web, coming from FML code, able to receive execution request via HTTP using input data and generating output data coded XML via SOAP protocol. This approach allows to realize a network of fuzzy controller, distributed around the world, usable from different hardware and different applications, as envisaged in Figure 5. Let's add a few comments about the standards that make this definition complete:

- SOAP (Simple Object Access Protocol) defines the internal structure of the XML documents that Web services consume and produce. SOAP is recognized as an industry standard and is widely adopted across many boundaries - across software vendors, hardware platforms, operating systems and programming languages. SOAP is one of the three cornerstones of today's Web services.
- A Web service lives at and is identified by its URI. This address is often called an endpoint. The identity has nothing to do with security in this context.
- A Web service carries its own description. It tells the outside world:
  1) what kinds of documents it exchanges (the interface definition)
  2) where the service lives (the URI - address)
  3) which transport protocols it can use for the exchange of documents (the binding).

  The language that we use for Web service description is called WSDL [19] - Web Services Definition Language. WSDL, an industry standard, is the second cornerstone of Web services. Please note that WSDL describes the Web service in its own context; it says nothing about orchestration with more Web services.
- A Web service has its home and the world wants to find where it lives to visit it. It needs a listing in an address directory. The industry standard, UDDI [21] (Universal Description, Discovery and Integration), is the third cornerstone of Web services. UDDI deals with registration and discovery of Web services. The UDDI registry is the yellow pages that the world uses to discover Web services.

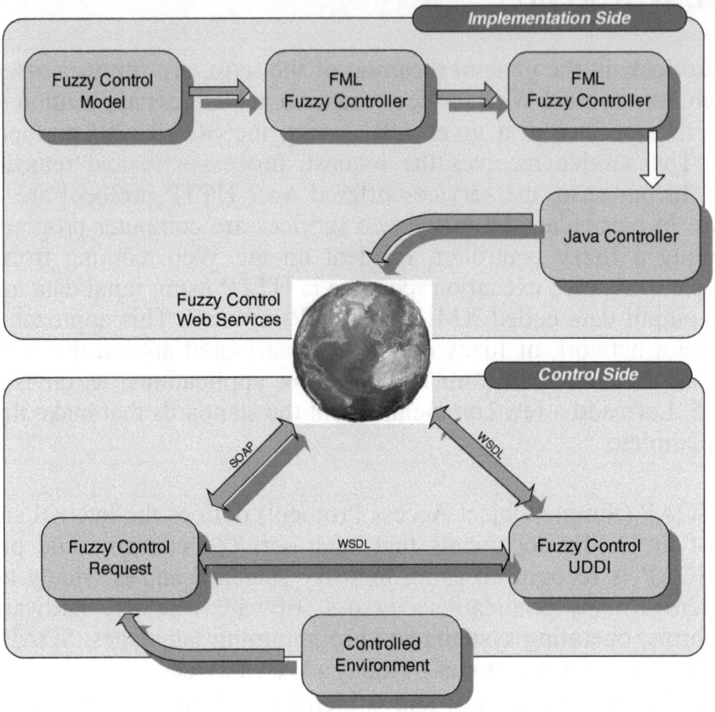

Figure 5 - FML script preprocessing

Putting it all together, a Web service is an entity that exchanges SOAP documents with the world, lives at its URI, is described by a WSDL document and can be listed and discovered in a UDDI registry. In particular, in the case of Fuzzy Web Services the SOAP messages contain the information about input and output controller. The URI of web service controller is obtained from attributes related FML <FuzzyControl> tag.

In fact, if our FML program uses

<**FuzzyControl** *name* = **"HVACController"** *ip* = **"192.168.4.12"**>

as openining tag, then the corresponding fuzzy web service will reside on: **http://192.168.4.12/HVACController.**

The WSDL of fuzzy web services describe the set of controlled and controlling information, in order to allows the fuzzy control request to retrieve from Fuzzy UDDI yellow pages the URI about the fuzzy controller web services appropriate for information contained in WSDL. The code implementing the fuzzy web services is obtained by using the XML XSLT technology as described in [18]. In the follow, the FML Web Services will be analyzed from two perspective: provider and requester.

5. FML Web Services Provider

The FML web services provider is any provider of one or more web services applied to Ambient Intelligence environments controlled by fuzzy logic systems. In order to realize an FML web services provider, a development plan has to be defined. The typical development plan for an FML web service provider consists of the following steps [22]:

1. develop the core functionality of the service.
2. develop a service wrapper to the core functionality (this could be an XML-RPC or a SOAP service wrapper).
3. provide a service description (a WSDL file in case of a SOAP application,).
4. deploy the service (by installing and running a standalone server or integrating with an existing web server).
5. publish the existence and specifications of the new service (by means publishing data to a global/private UDDI directory).

A snapshot of the service provider perspective is provided in Figure 6. Steps 1-3 can be accomplished using the XML technologies, and in particular, the XSLT technology. In fact, it is simple for XSLT module to receive in input a FML program and to generate the opportune output representing the different view of web services functionalities as shown in Figure 7.

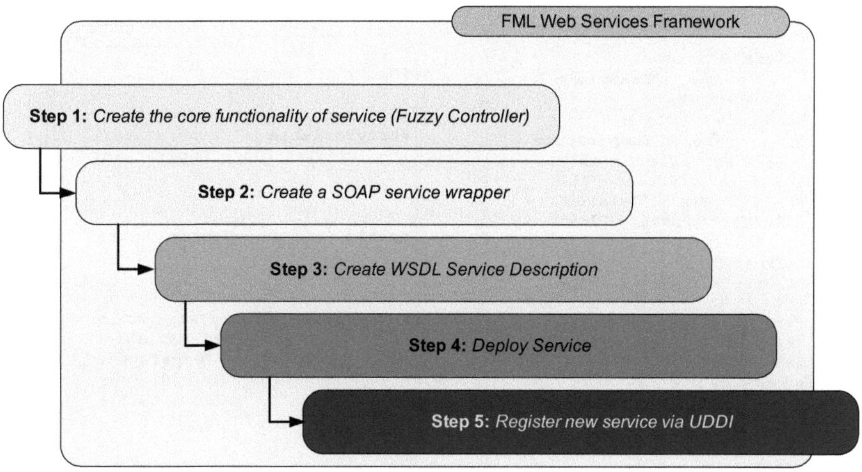

Figure 6 - Web Services Framework

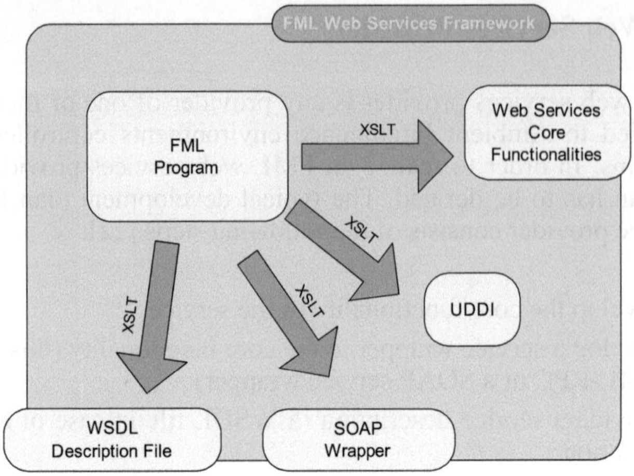

Figure 7 - FML Web Services Framework

Core Functionalities

In the case of FML web services the core functionalities are represented from the execution instances of fuzzy controllers modelled in FML language. The FML core functionalities are automatically developed by using the XML-XSLT technology able to translate the Fuzzy Markup Language code into executable computer program:

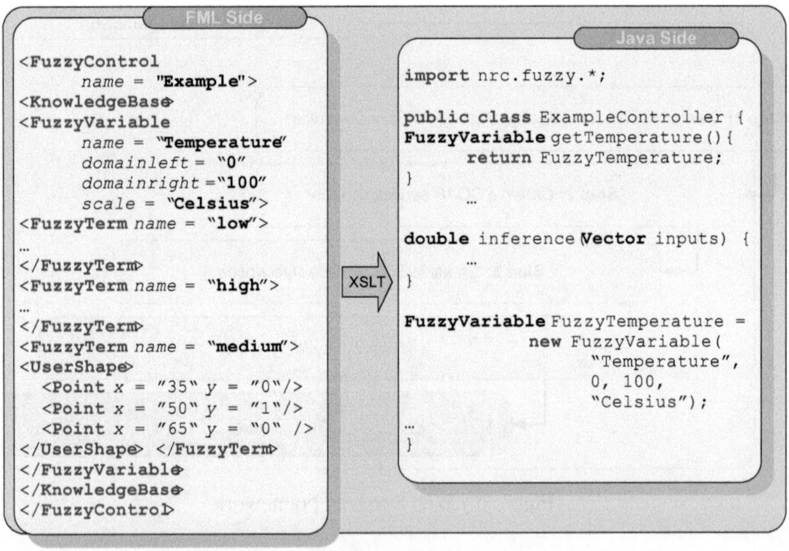

Figure 8 - FML Web Service Core Functionality Creation

The XSLT translator generates a Java class file able to instantiate the different controller parts composing the FOM and to expose, toward Ambient Intelligence applications, different methods, in particular:

- a set of access methods able to retrieve information about fuzzy controller components (FOM);
- an inference method able to apply the inference operators using the inputs sent from FML web services clients.

For instance, considering a fuzzy system able to control the lightening level using the luminosity sensor and time information, the generated methods are:

- **double** *inference*(**double** lux, **double** time);
- **FuzzyVariable** *getFuzzyLuminosity*();
- **FuzzyVariable** *getFuzzyTime*();

SOAP Service Wrapper

SOAP is an XML-based protocol for exchanging information between computers. Although SOAP can be used in a variety of messaging systems, and can be delivered via a variety of transport protocols, the main focus of SOAP is RPCs transported via HTTP. SOAP is platform-independent and therefore enables diverse applications to communicate. SOAP can be considered the core of application-to-application Ambient Intelligence framework, in fact, it allows the intercommunication beetween the controlled environment and the controller software. To gain a high-level understanding of SOAP, let's revisit our simple fuzzy service. The SOAP wrapper can be derived from the FML program considering the knowledge base part and, in particular, focusing the attention on the fuzzy variables set. Here is a sample SOAP request (HTTP headers omitted) realized translating an FML program containing two input variables as *indoorLuminosity* and *outDoorLuminosity*, based on the same universe of discourse (0-2000lux):

```
<?xml version='1.0' encoding='UTF-8'?>
<SOAP-ENV:Envelope
xmlns:SOAP-ENV="http://www.w3.org/2001/09/soap-envelope/"
xmlns:xsi="http://www.w3.org/2001/XMLSchema-instance"
xmlns:xsd="http://www.w3.org/2001/XMLSchema">
<SOAP-ENV:Body>
<ns1:inference
xmlns:ns1="urn:fml:LuminosityController"
SOAP-ENV:encodingStyle="http://www.w3.org/2001/09/soap-encoding/">
<indoorLuminosity xsi:type="xsd:double">1200</indoorLuminosity>
```

```
<outdoorLuminosity xsi:type="xsd:double">2000</outdoorLuminosity>
<timeOfDay xsi:type="xsd:double">1200</timeOfDay>
</ns1:luminosityController>
</SOAP-ENV:Body>
</SOAP-ENV:Envelope>
```

In analogous way, considering the output fuzzy variable of FML program and applying the XSLT tool it is possible to define the FML SOAP Response:

```
<?xml version='1.0' encoding='UTF-8'?>
<SOAP-ENV:Envelope
xmlns:SOAP-ENV="http://schemas.xmlsoap.org/soap/envelope/"
xmlns:xsi="http://www.w3.org/2001/XMLSchema-instance"
xmlns:xsd="http://www.w3.org/2001/XMLSchema">
<SOAP-ENV:Body>
<ns1:getLumResponse
xmlns:ns1="urn:fml:LuminosityController"
SOAP-ENV:encodingStyle="http://schemas.xmlsoap.org/soap/encoding/">
<return xsi:type="xsd:float">710.0</return>
</ns1:getLumResponse>
</SOAP-ENV:Body>
</SOAP-ENV:Envelope>
```

WSDL Service Description

WSDL stands for Web Services Description Language. It is an XML grammar able to describe and locate web services. In the case of fuzzy web services derived from FML program, the WSDL have to describe different information coming from different part of the FML program, and in particular: the name of service derived from the name attribute of fuzzycontrol tag; the access methods set offered by services derived from the FOM; the inference method. WSDL describes the data used as input parameters and return value from the FML web services. Obviously, the inference method uses double values both input parameters and return value, whereas, the access methods uses only the return value whose type is depending from the particular required FOM component. For instance, the WSDL file has to manage information as fuzzy variables and fuzzy rules in order to (remotely) access to such data.

UDDI– Publishing and Finding FML Web Services

The last step to implement the FML web services framework is the publish the obtained FML web controllers. The web services tool delegate to this scope is UDDI. UDDI is a technical specification for building a distributed directory of businesses and web services. Data is stored within a specific XML format, obtained from XSLT markup language. Moreover, UDDI specification includes API details for searching existing data and publishing new data. UDDI allows to manipulate three different catalog of information: the *white pages*, the *yellow pages* and the *green pages*. The last category contains technical information about a web service. Generally, this includes a pointer to an external specification and an address for invoking the web service. UDDI is not restricted to describing web services based on SOAP.

Concluding Remarks

Today there are evermore smart devices that adapt to our needs. These objects, connected by protocols and intelligent service brokering realize the modern AmI vision: smart devices should be able to (re)configurate the associations among human actions and their functionality in order to minimize and anticipate human intervention. To face with this problem is needed to conceive a middleware that should be able to: 1)connect the smart devices in the AmI environment; 2)allow them to interact by exchanging non only data but tasks; 3)adapt itself to the dynamically changing device characteristics. Our work proposes FML and its derivative, as basic component of the middleware. Therefore, inter-layer interactions are essential.

Throughout this program proper architectures for such systems will be developed. These architectures are not necessarily based on the well-known client/server paradigm but rather follow the peer-to-peer principle. This requires new structuring principles for services and applications that especially address the high requirements considering flexibility.

References

1. Basten, T., Geilen, M., de Groot, H. (2003) Ambient Intelligence: Impact on Embedded System Design. Kluwer Academic Publishers, Boston, 2003.
2. Holmquist L.E., Gellersen H.W., Kortuem G., Antifakos S., Michahelles F., Schiele B., Beigl, M., Maze R. (2004) Building intelligent environments with Smart-Its. IEEE Computer Graphics and Applications 24:56-64.
3. Geppert L. (2001) The new chips on the block . IEEE Spectrum 38:66-68.
4. Huns, M. N., Sehadri, S. (2000) Sensor + Agent + Networks =Aware Agents, IEEE Internet Computing 84-86.
5. Sun Microsystems, "Jini architecture specification", http://www.sun/com/jini/specs/
6. Microsoft, "Universal plug and play", http://www.upnp.org

7. Munguia Tapia E., Intille S.S., and Larson K.(2004) Activity Recognition in the Home Using Simple and Ubiquitous Sensors In: A. Ferscha, F. Mattern(eds) Proceedings of PERVASIVE 2004 Berlin Heidelberg: Springer-Verlag, 2004, pp. 158-175
8. Beaudin J., Intille S. and E. Munguia Tapia.E., Lessons Learned Using Ubiquitous Sensors for Data Collection in Real Homes (2004) In: Proceedings of Extended Abstracts: CHI 2004 Connect: Conference on Human Factors in Computing Systems, ACM Press
9. Strohbach M., Kortuem G., Gellersen H-W. , Kray C. (2004) Using Cooperative Artefacts as Basis for Activity Recognition. In: Proceedings of 2nd European Symposium on Ambient Intelligence (EUSAI), Eindhoven, The Netherlands, pp.49-60.
10. Polat E., Yeasin M., Sharma R. (2003) Robust tracking of human body parts for collaborative human computer interaction. J of Computer Vision and Image Understanding 89: 44-69 .
11. Wren, C.R., Azarbayejani, A., Darrell, T., Pentland, A.P., Pfinder: real-time tracking of the human body (1997). IEEE Transactions on Pattern Analysis and Machine Intelligence 19:780-785.
12. Zhao T., Nevatia R (2004) Tracking multiple humans in complex situations. IEEE Transactions on Pattern Analysis and Machine Intelligence 26:1208 – 1221.
13. Lesser V., Ortiz C., Tambe M. (2003.) Distributed Sensor Networks: A Multiagent Perspective, volume 9. Kluwer Academic Publishers. 2003
14. Kurihara S., Fukuda K., Hirotsu T. , Aoyagi S., Takada T., Sugawara T. (2004) Multi-agent Framework for Human-Environment Interaction in Ubiquitous Environment, Workshop on Agents for Ubiquitous Computing, New York City.
15. O'Hare G.M.P., O'Grady M.J., Collier R., Keegan S., Kane D.O, Tynan R., Marsh D. (2005) Ambient Intelligence Through Agile Agents. In Ambient Intelligence for Scientific Discovery, Lecture Notes in Computer Science, Volume 3345, pp 286-310.
16. "Extensible Markup Language (XML) 1.0 III Edition", W3C Recommendation, http://www.w3.org/TR/2004/REC-xml-20040204/
17. Acampora G., Loia V. (2004) Fuzzy Control Interoperability for Adaptive Domotic Framework. In: Proceedings of 2nd IEEE International Conference on Industrial Informatics pp 184-189.
18. Acampora G., Loia V. (2005) Fuzzy Control Interoperability and Scalability for Adaptive Domotic Framework. IEEE Transactions on Industrial Informatics 1:97-111
19. W3C, Web Services Description Language (WSDL) Version 2.0 Part 1: Core Language, http://www.w3.org/TR/wsdl20/
20. W3C, SOAP version 1.2, http://www.w3.org/TR/soap/
21. uddi.org, http://uddi.org/pubs/DataStructure_v2.htm
22. Cerami E. (2002) Web Services Essential, O'Reilly Ed.

Development of Ontologies by the Lowest Common Abstraction of Terms Using Fuzzy Hypernym Chains

Rafal A. Angryk[1], Jacob Dolan[1], and Frederick E. Petry[2]

[1]*Department of Computer Science, Montana State University*

Bozeman, MT 59717-3880, USA

[2]*Naval Research Laboratory, Mapping, Charting and Geodesy*

Stennis Space Center, MS 39529, USA

Abstract

We present an automated approach to produce the hierarchy of abstracts for a set of concepts inserted as an input. Our goal is to provide a tool allowing generalization of terms without the necessity of being an expert in the domain to be generalized. Through the use of ontologies, a set of terms can be automatically generalized into the next minimally abstract level in a concept hierarchy. The algorithm for discovering these minimally abstract generalizations and the unsupervised construction of a fuzzy concept hierarchy is presented. The algorithm presented is a prototype and will be followed by future research concentrated on efficiency improvements.

1 Introduction

We describe an approach that allows for automatic creation of ontologies understood as taxonomies of abstract terms based on the list of descriptors provided by the user. This approach can be utilized in many applications including: (1) creation of general users profiles based on their actions (e.g. http://www.amazon.com), (2) generalizations of search terms commonly

R.A. Angryk et al.: *Development of Ontolgoies by the Lowest Common Abstraction of Terms Using Fuzzy Hypernym Chains*, StudFuzz **204**, 123–148 (2006)
www.springerlink.com

used in customer self-help systems (i.e. FAQs), and also (3) generation of concept hierarchies applicable for automatic summarization of large sets of descriptive data in data mining applications (e.g. DBMiner http://www.dbminer.com). Here we make use of a specific popular online lexical reference system, WordNet [1] as a common source of information, but we believe that the proposed algorithm allows the approach to find applications in many other areas.

In this chapter we will investigate the possibility of automatic creation of fuzzy concept hierarchies, without the necessity of a user being an expert, or being forced to employ one, in the domain she needs to have her terms generalized. We will focus on automation of the whole process of finding abstracts and on the employment of a freely available ontology system (i.e. WordNet). Hypernyms (generalizations) extracted by WordNet and fuzzy concept hierarchies are used in order to find the user's categories of interest. The user's categories of interest are computed based on a novel approach using fuzzy concept hierarchies (FCH) extracted from the WordNet ontology. We will also discuss crucial components of our approach and the motivation behind choices made during implementation of the algorithm. Finally we present our algorithm and discuss its elements, which are crucial for the successful implementation and provide a short example of its functioning.

2 Background

2.1 Types of Hierarchies and their Applicability for Abstract Concept Representation

A concept hierarchy (also referred to as a generalization hierarchy, or as a hierarchy of abstracts), as defined by Hamilton et al. [2], is a rooted tree that reflects a taxonomy of generalization concepts. Usually the single, most abstract concept (e.g. ENTITY) is placed as the root of the generalization hierarchy, and the leaves (i.e. concepts at the lowest abstraction level) contain the original values from the user.

A generalization hierarchy can either be overlapping or disjoint. In an overlapping, also known as fuzzy, concept hierarchy terms at the lower level of abstraction can be assigned to multiple parent nodes (i.e. abstracts). In a disjoint (i.e. crisp) concept hierarchy, each term can have only one abstract at the next level of abstraction. An example of such a concept hierarchy characterizing a generalization of students' status ({master of art, master of science, doctorate} $\subset$ graduate, {freshman, sophomore, junior, senior} $\subset$ undergraduate, {undergraduate, graduate} $\subset$ ANY) is presented in Figure 1, which appeared in [3].

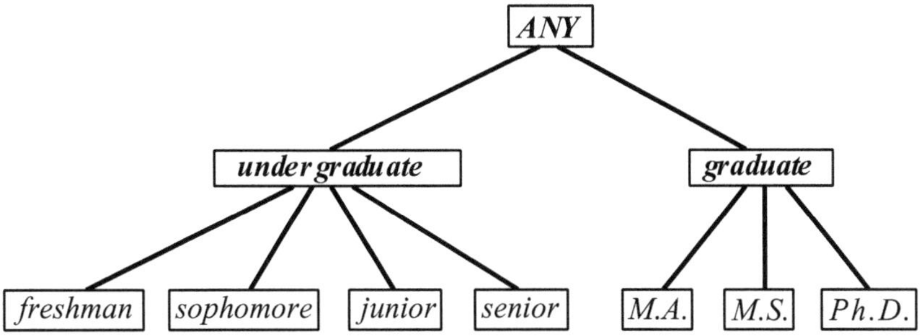

Fig. 1 A concept tree for attribute STUDENT_STATUS [3]

There has been a significant amount of work devoted to utilization of crisp concept hierarchies. Han et al. [3] used them to summarize large databases for data mining purposes. The work initiated by Han and his co-researchers [4-9] was extended further by Hamilton, Hilderman with their co-workers [2, 10-11], and also by Hwang and Fu in [12]. The hierarchical summarization methods have been implemented in commercially used systems (e.g. DBLearn/DB-Miner [13]), and applied to a number of research and commercial databases to produce interesting and useful results [2].

A crisp approach however does not handle the problem of overlapping meanings or senses, when the transformation from lower level of concepts to the more abstract level is performed (certain descriptors entered by the user may fit well into more than one of the concepts placed at the immediately higher level). Trivial examples of such cases can be found everywhere, e.g. how to generalize three colors, such as black, gray and white to two abstract terms: dark, light, or to which continent we should assign the Russian Federation, taking under consideration an undeniable fact that almost one fourth of this country lies west of the Urals.

The non-crisp nature of the world naturally leads to utilization of a fuzzy approach, where a single lower-level term could have partial membership in more than only one higher-level concept. Fuzzy concept hierarchies seem to better capture human approaches to induction. Typically the more abstract the level of concepts to be used in data generalization, the less certain experts are in definite assignment of particular lower-level concepts to them. Utilization of the partial membership concept allows for more natural and accurate modeling of real-life circumstances.

A number of approaches using hierarchies for fuzzy summarization have been developed. Lee and Kim [14] used fuzzy IS-A hierarchies to generalize database records to more abstract concepts, where IS-A relation was used to reflect a popular in ontologies "kind-of" dependency. Lee [15] applied fuzzy generalization hierarchies to mine generalized fuzzy quantitative association rules. Cubero et al. [16] introduced fuzzy gradual rules for data summarization. Some of our work on fuzzy hierarchies of concepts can be found in [17-20].

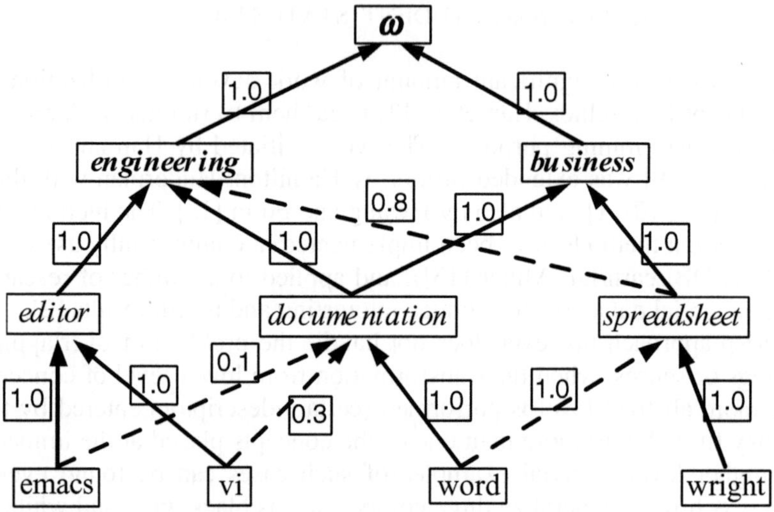

Fig. 2 A fuzzy ISA hierarchy on computer programs [14]

In this work we focus on the utilization of fuzzy hierarchies and semantic relations between concepts derived from WordNet for the purpose of achieving a more flexible representation of dependencies occurring during generalization of multiple terms.

2.2 WordNet as the Source of Ontology

To understand our approach for automated creation of fuzzy hierarchies of terms generalizing the originally given data, we will describe details of WordNet here. WordNet [1, 21-25] is an on-line lexical reference system which links English verbs, nouns and adverbs to groups of synonyms, which are further linked via semantic relations allowing for understanding of the words' meanings. It was originally developed by a group of psychologists and linguists from the Cognitive Science Laboratory at the

Princeton University with the purpose of searching dictionaries conceptually rather than alphabetically [24]. The power of searching dictionaries conceptually is critical for our approach to concepts' generalization.

WordNet currently has a database of over 200,000 word-sense pairs that can be used to construct generalization hierarchies. Our approach is based on the collections of terms stored in WordNet and the semantic relationships among terms that the system is capable of providing. The following semantic relations are included in the WordNet database: Synonymy, Antonymy, Hyponymy (Hypernymy), Meronymy, Troponomy, and Entailment. Hyponym and hypernym chains are the main topics of interest in our current research as they can provide sufficient information to form fuzzy concept hierarchies of abstracts. We are specifically interested in the ability of WordNet to produce a hypernym chain for almost every English word. WordNet is likely the largest collection of linguistic information freely available on the Internet, so for our research it provided enough linguistic information to produce satisfactory results for generalization of English terms.

A hypernym is a superordinate word. More specifically, a word's hypernyms are more general concepts of that word. So a hypernym chain for a specific word is an ordered collection of terms representing the original term at different levels of abstraction. The hypernym chain reflects a single sense of the word, and it can be interpreted as conjunction of IS-A relationships. Each word has an IS-A relationship with it's abstract, and then this abstract has an IS-A relationship with its parent, and so on. For instance, for a word such as BURGUNDY, WordNet provides multiple hypernym chains, which can be represented in a Fuzzy Concept Hierarchy (see Fig. 3). The complete hypernym hierarchy for the word burgundy (see Fig.1) includes the hypernym chains for each of the burgundy's senses (meanings). For "burgundy" specifically, there are three senses, which provide a total of twenty-four possible generalizations for the word burgundy.

Since there is normally a superordinate for every term entered by the user, the semantic relation of generalization generates a hierarchical structure where hyponyms (i.e. lower level terms) are placed below their superordinates (i.e. hypernyms). The hyponym inherits all features of its abstract and adds at least a single feature that distinguishes it from its' superordinate and from other hyponyms of that superordinate. Hypernym is an abstract term used to designate a whole class of specific terms.

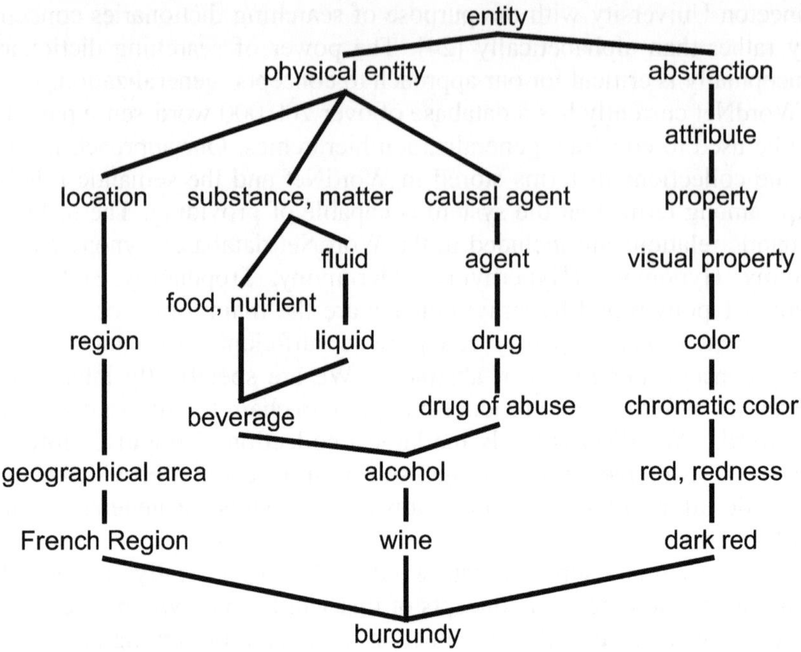

Fig. 3 Set of hypernym chains for BURGUNDY presented in the form of a Fuzzy Concept Hierarchy

It is important to observe the transitive and asymmetrical character of a generalization relation (e.g. Wine IS-A (KIND-OF) Alcohol). If a given word is an abstraction of a particular term, it is also going to be an abstraction of this term's specializer. The farther we are from the original word, the more abstract the hypernyms become just as in a concept hierarchy. At the end of the hypernym chain, which is the top of the concept's generalization hierarchy, the abstract term for an original word is always the farthest generalization of the original word that still describes that original word.

The fact that most words have multiple senses complicates the use of WordNet hypernym ontologies for concept generalization. A word's generalization hierarchy loses its crisp character. Without having additional knowledge about the terms that need to be generalized we cannot clearly specify which abstract to choose in order to produce a generalization hierarchy for the original terms. We use some "support" based on the observation that a group of inserted concepts may suggest the appropriate sense path for generalization. Given a set of concepts to generalize, we generalize the set of the semantically closest abstracts of those concepts; in other words, the generalization should occur at the least abstract level possible.

The task becomes even more complex, when we deal with the generalization of concepts that have multiple senses which overlap. Given a set of terms such as *cherry*, *burgundy*, and *lemon*, the terms' senses have many common abstracts. *Cherry* and *lemon* are both types of *fruit*, and *burgundy* is a type of *wine* that is related to *fruit* through the abstract concept of *substance*. They are all also related as *cherry* and *burgundy* are types of the color *red*, and *lemon* is a type of the color *yellow*. Both *red* and *yellow* are types of *chromatic color*. Semantically, the three terms are all closer to the abstract concept of *chromatic color* than they are to the more abstract concept of *substance*. However, the conclusion that *substance* is more abstract than *chromatic color* could not be derived without human reasoning. Without background knowledge about the generalized terms, it seems more accurate to choose an abstract concept that is semantically closest to all of the terms. The greater the number of distinct values to generalize, the more likely an accurate abstraction can be achieved through this process which is the major rationale for our generalization algorithm .

Some measures of the semantic closeness of concepts in taxonomies have been developed. Calmet and Daemi [26-27] extend Shannon's measure of entropy to measure the distance between two concepts X and Z by measuring the reduction of uncertainty in X through the knowledge of a third concept Y, when Z is known. In the following let X be a random variable with a probability mass function $p(x) = Pr\{X = x\}, x \in \Omega$ and alphabet Ω; $p(z) = Pr\{Z = z\}, z \in \Psi$ and alphabet Ψ:

$$d(X,Z:Y) = H(X|Z) - H(X|Y,Z) \text{ where } H(X|Z) = \sum_{x\in\Omega} \sum_{z\in\Psi} p(x,z) log\ p(x|z)$$

As Calmet and Daemi require a method of determining the probability of a term to measure the entropy, their method fails to satisfy our restriction of no background knowledge. However, their evaluation of entropy as a measure of information is potentially a very interesting approach to the problem.

Resnik [28] proposed that the distance (similarity) between two concepts could be measured by the amount of information that the concepts shared as:

$$sim(c_1,c_2) = \max_{c\in S(c_1,c_2)} [-log\ p(c)] \text{ where } p(c) = \frac{freq(c)}{N}$$

As with the measure of entropy above, this method relies on the frequency, $freq(c)$, of concepts to determine their probability which is employed to measure the similarity of the concepts' information content. However, this does not provide a solution for our problem when multiple senses of a concept are present as the frequency measure does not discriminate between the individual word-senses of a concept. In the past, the frequency of a word-sense was available through a semantic concordance in WordNet, but current versions of WordNet no longer maintain this concordance. Other methods measures have been explored and evaluated [29] to provide a measure for semantic similarity. All of these proposed solutions either require background knowledge (often in the form of a probability measure) to measure semantic similarity between two concepts and their shared common abstracts or they do not allow the discrimination between the different senses of concept.

3 Desirable Properties of Concept Hierarchies

Before describing details of the algorithm, we need to discuss fundamental requirements we would like our concept hierarchies to maintain and their influence on the behavior of our algorithm as well as on the properties of generated concept hierarchies. Here are the requirements we would like our generalization hierarchies to satisfy:

Maintaining Information about All Terms Inserted by the User

Our concept hierarchy needs to consistently reflect the information about all originally given terms at every level of the hierarchy/abstraction. In other words we want to make sure that as we move to a higher level of abstraction, we still cover the space of senses intended by all original terms. Formally:

$$SpaceOfSenses(C^{\ell+1}) \geq SpaceOfSenses(C^{\ell}) \text{ for every } \ell \text{ such that } 0 < \ell + 1 \leq H \tag{1}$$

where C^{ℓ} is a set of concepts $\{c_1^{\ell}, c_2^{\ell}, ..., c_n^{\ell}\}$ at the lth abstraction level, and the level is in the range *<0, H>*. Also, *H* is total height of concept hierarchy, *0* corresponds to 0th abstraction level (i.e. terms originally inserted by the user), and the increment of abstraction is *1*.

At a higher level of abstraction we usually deal with more general terms, which have broader meaning than the lower level concepts they generalize. However, our goal here is to guarantee that at every new level, the abstracts cover <u>at least</u> the same space of senses as the original terms.

This property is strongly related to the strategy of *Vote Propagation* introduced by Han in [3, 9]. All of the original database attribute values, which are generalized, need to maintain their full representation (which is more and more abstract) at every level of concept hierarchy; at the same time, to avoid confusing duplications, we do not allow any of the original concepts to be represented more than once at each level.

Since each of the generalized concepts can have more than one meaning, we needed to somehow transfer this requirement to the level of the concepts' senses. The task leads to the important dilemma. From one side we needed to take under consideration the fact that originally given concepts create separate entities from the user's point of view, and should not be treated differently based on the vagueness (i.e. multi-meanings) they may, or may not carry. At the same time, concepts having multiple meanings should have all of these generalized, since during the automatic generalization we are not able to distinguish between meaningful and meaningless senses from the user's or task's perspective.

As we have already discussed, multiple senses of terms can be modeled with a Fuzzy Concept Hierarchy. This type of hierarchy can reflect the relations that have a many-to-many character, where a single concept c can have more than one direct abstract c_a, and c can belong to the higher-level concept c_a to a certain degree, μ_{cc_a}, For example, since the word *burgundy* has three direct abstract descriptors: *French region*, *wine*, and *dark red* we can split the memberships degrees evenly: $\mu_{burgundy\text{-}French\ region} = 0.33$, $\mu_{burgundy\text{-}wine}=0.33$; and $\mu_{burgundy\text{-}dark\ red}=0.33$. The even, linear distribution of memberships reflects the situation where we do not have preferences concerning any of the senses extracted from WordNet. At the same time, due to the fact that all fractions of membership sum to unity, we are assured that each of the concepts is represented with the same validity as the others.

The vote of each concept, split evenly amongst each sense of a concept, needs to be propagated upwards in the tree so that at any level each concept vote is fully represented. Consider a possible next step of generalization according to the extracted hypernym chain, where the descriptor *French region* has only one direct abstract (i.e. *geographical area*). In order to preserve completeness of the fuzzy model, we assign exactly the same weight to the $\mu_{French\ region\text{–}geographical\ area}$ as to the link incoming to concept French region: $\mu_{French\ region\text{–}geographical\ area}=0.33$ since the preceding membership value in the analyzed generalization path (i.e. $\mu_{burgundy\text{-}French\ region}$) is 0.33. If branching occurs, as it does after the concept *alcohol*, we have to evenly split the preceding membership value in order to preserve completeness: $\mu_{alcohol\text{-}drug\ of\ abuse}=0.16$, and $\mu_{alcohol\text{-}beverage}=0.16$, as shown in Figure 4.

Formalizing this characterization of the proposed membership function, we can define the two following properties as necessary for our concept hierarchies:

1. Denote the original term by c^0 (a single concept at the 0-abstraction level, i.e. placed at the bottom of the FCH) and all of its direct abstracts by a set at the first level of abstraction: $C^1=\{c_1^1, c_2^1, ...\}$; then we have to preserve the following condition in order to keep the model's consistency at the first step of generalization:

$$\sum_{i=1}^{\|C^1\|} \mu_{c^0 c_i^1} = 1.0 \qquad (2)$$

In other words, the sum of weights assigned to the links outgoing from every original term has to be equal to the unity at every level of abstraction.

2. Denote a single node in the FCH as $c_p^k \in C^k$, where k symbolizes the level of abstraction (we assumed upward numeration of abstraction levels in the FCH), and C^k symbolizes a set of all abstract concepts at the given k-abstraction level; then we have to assure this condition in order to maintain generalization completeness:

$$\sum_{i=1}^{\|C^{k-1}\|} \mu_{c_i^{k-1} c_p^k} = \sum_{j=1}^{\|C^{k+1}\|} \mu_{c_p^k c_j^{k+1}} \text{ for every } k > 0 \text{ and } p \le \|C^k\| \qquad (3)$$

In other words, the sum of weights from all links incoming to each node in the FCH has to be equal to the sum of weights in all outgoing edges.

The rationale behind this approach is quite obvious: when the sum of all memberships derived during generalization of the concept burgundy is equal to 1.0 at each level of the concept hierarchy (at each generalization step), the completeness of the model is preserved. This guarantees the word *burgundy* will be constantly represented as a single concept at each level of abstraction. The complete fuzzy generalization hierarchy of the concept *burgundy* is presented in the Figure 4.

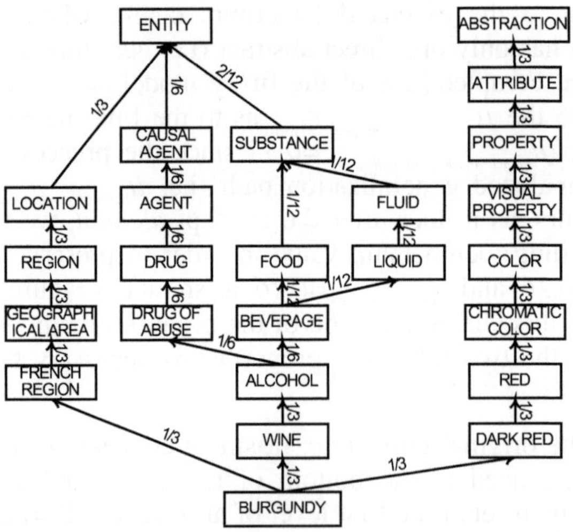

Fig. 4 Fuzzy Generalization Hierarchy for the concept burgundy extracted from the hypernym chains

Detailed Abstraction Representation of Original Terms

Without expert knowledge for generalization we cannot definitely declare which hypernyms and which abstraction levels are the most appropriate for generalization of the user's terms. What makes this even more complicated is the fact that terms occurring in natural language do not always appear at equivalent intervals of abstraction (e.g. The Vatican is a place which is considered to be a city and a country at the same time, whereas the distinction between city and country abstraction levels is much more significant when other names of locations are generalized). So we are not always guaranteed to obtain appropriate abstract terms at every level of a generated hierarchy. Moreover, during automatic generalization, we are not able to decide which of the senses represented by terms with multiple interpretations are the most appropriate from the user's perspective. We cannot decide which of the discovered abstraction levels should remain and which ones should be eliminated. To make sure that our hierarchy reflects all generalization possibilities, and assuming that a user can make her final decisions, we will employ a strategy used in attribute oriented induction [3, 9]. When analyzing strategies applicable for hierarchical generalization of large datasets, the minimal Concept Tree Ascension is an appropriate approach for avoiding the risk of overgeneralization. This strategy is linked directly to the issue of choosing an appropriate abstraction increment for every level of our concept hierarchy.

To insure that we avoid overgeneralization, we follow a minimal abstraction increment approach, in the sense that a new level of concept hierarchy is going to be generated each time new abstract merging at least two terms is discovered. Formally:

$$\ell_{next} = \ell_{current} + \text{k for every } \ell \in [0, H] \tag{4}$$

where k is Minimal Abstraction Increment Available, based on the list of terms provided for generalization, H is total height of concept hierarchy, and 0 corresponds to abstraction level of the originally inserted terms.

The requirement of a minimal abstraction increment has both positive and negative points. (1) One of the major drawbacks is that in the worst case, when the user inserts totally unrelated terms, it is possible that only two terms (or, to be more exact – two senses represented by these terms) will be generalized in every new level of the concept hierarchy. This would lead to generation of very high concept hierarchies (in the worst case, for *n* senses we may generate even *n-1* levels), which have dendogram-like character. (2) At the same time, we need to remember that more levels of abstraction provide an advantage since the user is given more detailed representation of original terms, and more thorough generalization. A fundamental motivation for this work is the assumption that our algorithm will be used to generalize data that is somehow related, so the case of dendogram-like generalization should occur rarely.

The challenge of maintaining consistency of all original senses through the entire generalization process, in addition to the minimal abstraction requirement, leads to the following possible cases during automatic generalization:

1. The ideal case of generalization occurs when an abstract term, c_a, is a total abstraction of two (or more) concepts from the lower abstraction level, denoted here as c_{10} and c_{20}. c_a can even be a set of common hypernyms occurring at the same abstraction level, denoted as $c_{a1}...c_{an}$ where $c_{a1}...c_{an}$ combined have full membership of both c_{10} and c_{20}. We would represent this as $c_{10} \subset c_a$ and $c_{20} \subset c_a$, as is shown in Figure 5 where $s_1(c_{10})$ and $s_1(c_{20})$ represent the senses of the two concepts.

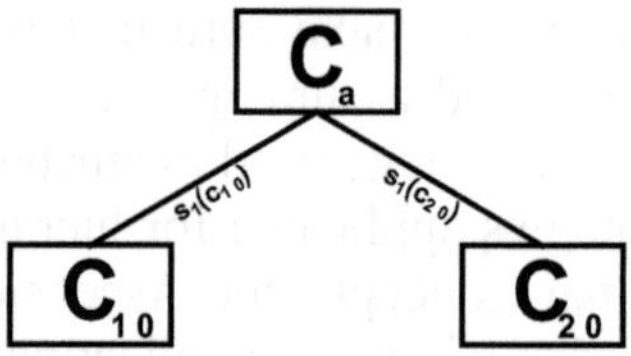

Fig. 5. c_a is a generalization with full membership both of c_{10} and c_{20}

An example of this case is the generalization of the concepts *canary yellow* and *old gold*. They will have a full membership with their common hypernym *yellow* as in Figure 6.

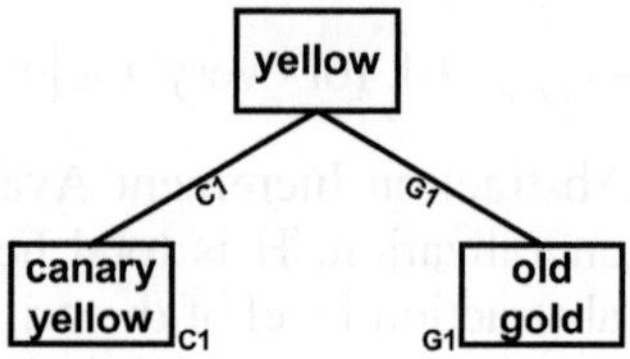

Fig. 6. Yellow is a generalization with full membership of canary yellow and old gold

2. In the second case we may have a common hypernym c_a that is an abstract for only some of the senses represented by its specializer concepts c_{10} and c_{20}. Assume c_a does not cover all senses represented by its specializer c_{10}. Then the remaining meanings of c_{10} must be still represented at the next abstraction level as a union of the senses represented by c_a and the remaining senses of c_{10}, denoted here as c_{11}. It is important to realize that in this case: $c_{11} \subset c_{10}$, but $c_{10} | c_{11} \neq \{\emptyset\}$, and that: $c_{10} \cap c_a \neq \{\emptyset\}$,

$c_{20} \cap c_a \neq \{\emptyset\}$, but $c_{11} \cap c_a = \{\emptyset\}$. The same analysis applies to c_{20} if all of its senses are not covered by c_a.

If, due to limitations of natural language (a limited number of terms and their vagueness), there is no appropriate descriptor of all senses of c_{10} not generalized to c_a, the name of c_{10} may itself need to be repeated at the new level, assuming it maintains a restricted meaning (only a subset of senses of c_{10}). In the case where c_{11} needs to be represented by the same term as c_{10}, it is considered to be a "temporary specializer" of c_{10}, since it has a restricted number of the senses provided by c_{10}, which are maintained only to be properly generalized at a higher abstraction level. This relationship is shown in Figure 7. Note that the figure uses a dotted line to signify the *"temporary specializer"* relationship and a solid line to denote a generalization and that $s_1(c_{10})$ and $s_2(c_{10})$ encompass the senses of the concept c_1.

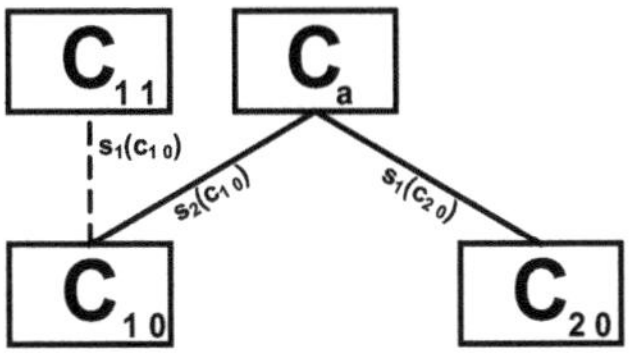

Fig. 7 c_a is a generalization of c_{10} and c_{20}. It has a partial membership of c_{10} and a full (or partial) membership of c_{20}. c_{11} has a partial membership of c_{10}

An example of this second relationship is the generalization of the concepts *lemon* and *cherry*. The common hypernyms *edible fruit*, *fruit tree*, and *chromatic color* do not fully represent all senses of *lemon* or *cherry*. For example another sense of cherry could be as a type of wood and for lemon, a distinctive flavor or a poorly- made car. Their unrepresented senses are then passed into the next level of the concept hierarchy as illustrated in Figure 8.

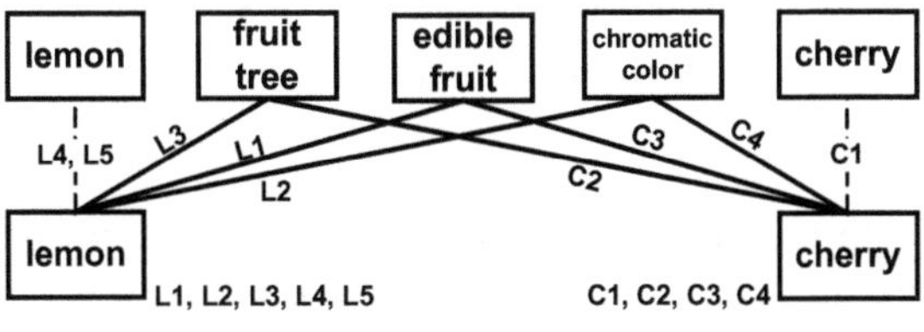

Fig. 8. *Cherry* and *lemon* must pass their unrepresented senses into the next level of the concept hierarchy

3. The third relationship exists when a concept c_{10} is a common hypernym of a concept c_{20}. It only arises when concepts of differing levels of abstraction are being generalized which should occur rarely as it is the result of improper data characterization (i.e. concepts that are of different levels of abstraction in the same attribute). c_{20} has a partial or full membership in c_{10} if senses characterized by c_{10} contain senses of c_{20} (i.e. c_{10} is actually a generalization of c_{20}). It is important to note $c_{20} \subset c_{11}$ and $c_{11} = c_{10}$. This relationship is shown in Figure 9.

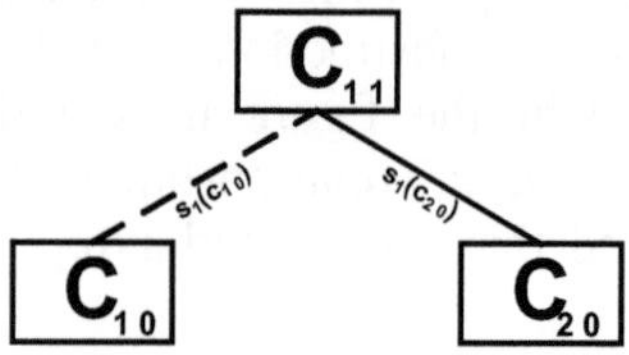

Fig. 9 c_{11} is a generalization of c_{20} and c_{10}. It has a partial (or full) membership of c_{10} and a partial (or full) membership of c_{20}

An example of the third relationship is the generalization of the concepts *red* and *burgundy*. *Red* is a generalization of *burgundy*, but it is not a generalization of itself or in other words, *red* is the 0th specialization of *red* and it is also the 0th generalization of *red*. *Red* must pass all of its unrepresented senses into the next level of the concept hierarchy even though it serves as a generalization of the concept *burgundy*. Note that although *red* is now represented by more than one concept in the next level, no actual generalization of *red* has yet occurred as shown in Figure 10.

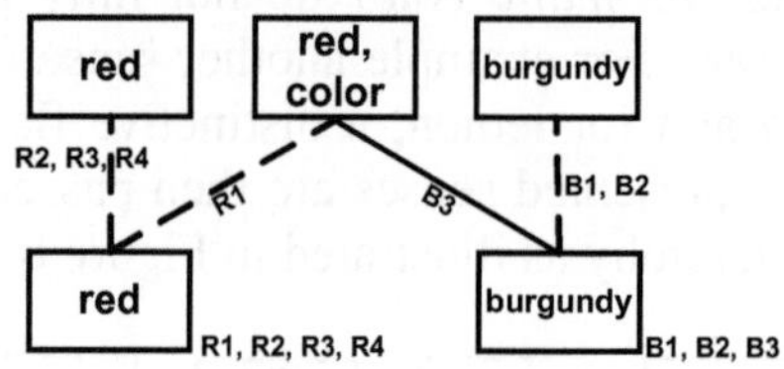

Fig. 10. *Burgundy* generalizes to a sense of *red* but *red* does not generalize to itself

4. The fourth relationship occurs when the two concepts c_{10} and c_{20} do not share any common hypernyms at the next level of abstraction, but they share common senses with other concepts, which were already generalized. c_{11} and c_{21} represent the senses of c_{10} and c_{20} that were not already merged with another concept's sense. This relationship is shown in Figure 11.

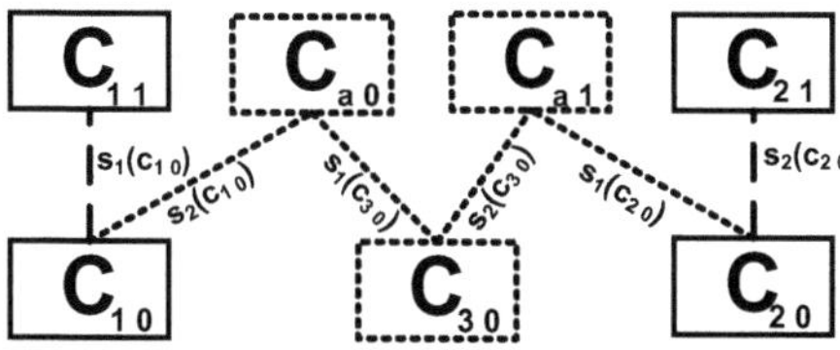

Fig. 11. c_{11} is not a generalization of c_{10}, but it does have a partial membership of c_{10}. c_{21} is not a generalization of c_{20}, but it does have a partial membership of c_{20}

An example of the fourth relationship can be illustrated by the generalization of the concepts *lemon*, *cherry* and *burgundy*. The common hypernyms of *lemon-cherry* and *cherry-burgundy* are found and placed into the concept hierarchy. However, generalizations of *lemon* and *burgundy* were found as well, but their abstraction level is much higher so they would not be included at the current generalization level. Some senses of both *lemon* and *burgundy* are passed into the next level of the tree without generalizing into any common hypernym between the two concepts as in Figure 12.

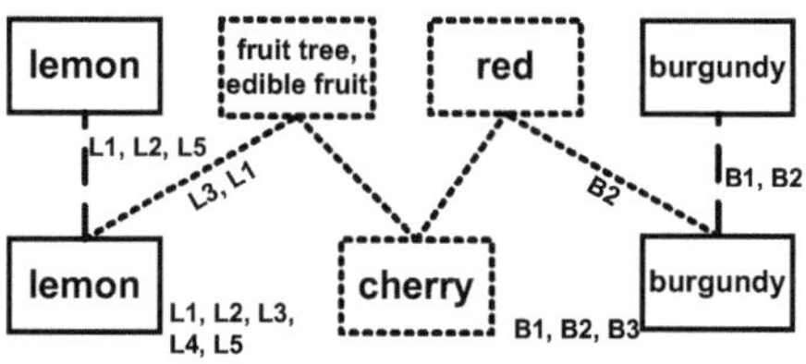

Fig. 12 *Lemon* and *Burgundy* share no common hypernym at the next level of abstraction so they must pass their unrepresented senses to the next level

In Figure 2, we displayed the resulting full fuzzy concept hierarchy for the concepts *cherry*, *burgundy* and *lemon*. The senses of each concept were traced through the tree to show where they are being represented at each level. Note the relationships that are displayed between each of the concepts. *Lemon* and *cherry* at level 0 have the second relationship where they have two common hypernyms but need to have some of their senses propagated to the next level in the tree to have full membership at that given level. *Lemon* and *burgundy* have only the fourth relationship at the level 0 of the tree. If *red* were one of the concepts, *burgundy* and *cherry* would have had the third relationship with *red* as *red* would be a very close hypernym to both of those concepts.

4 Building Hierarchies of Abstracts on Request with the Least Abstract Common Hypernym (LACH) Algorithm

4.1 Discovering Common Hypernyms and their Degree of Abstraction

The prototype that provides the automated generalization is designed using a level-based fuzzy concept hierarchy, T, built in a bottom-to-top approach. At the base of the hierarchy, T_0, each node is an original concept to be generalized into a more abstract concept. At each level of the hierarchy, T_k where $k > 0$, there exists at least one node where two concepts are being generalized into a more abstract concept.

As we climb the concept hierarchy, we can find the least abstract common hypernyms to generalize the original concepts on a level-by-level basis. The algorithm discovers the common hypernyms of two concepts, c_1 and c_2, by merging their senses' hypernym chains. The hypernym at which a sense from c_1 merges with a sense from c_2 is an *abstract common hypernym*, c_a, for both concepts. Any two concepts may have zero or more common hypernyms. Each c_a is a candidate to be placed into the fuzzy concept hierarchy. Each of the common hypernyms must be assigned a measure to determine the *degree of abstraction*, $c_a.d$, in the context of the concepts it generalizes.

The algorithm measures the *distance of abstraction*, d_{cc_a}, between a concept c and its generalization c_a by finding the minimum path length leading from c to its abstract concept c_a (measuring for each sense of c) and weighting it by a *degree of vagueness* of generalization from c to c_a, which is the vote of c's senses (i.e. hyperym chains) that lead to c_a. A concept's distance of abstraction to itself, d_{cc}, equals 0 as there is a 0 edge count between the concept and a set of senses of that given concept.

In the following, Ω_c represents the set of senses in the concept c, where s is a sense such that $s \in \Omega_c$. If $s \in \Omega_{c_a}$ we denote this fact as $\mu(s, c_a) = 1$, otherwise if $s \notin \Omega_{c_a}$, $\mu(s, c_a) = 0$; S_{cc_a} is the set of all paths, representing senses, leading from c to c_a, that is $S_{cc_a} \subset \Omega_{c_a}$ We can calculate the *degree of vagueness,* v_{cc_a}, using the following formula:

$$v_{cc_a} = 1 - \frac{\sum_{s \in \Omega_c}^{\|\Omega_c\|} \mu(s, c_a)}{\|\Omega_c\|} \tag{5}$$

The sum of memberships $\mu(s, c_a)$ for every $s \in \Omega_c$ is subtracted from 1 in order for v_{cc_a} to be proportional to the distance of abstraction d_{cc_a}, which is calculated as follows:

$$d_{cc_a} = v_{cc_a} \cdot Min[S_{cc_a}] \tag{6}$$

where $Min[S_{cc_a}]$ denotes the shortest generalization path leading from c to c_a.

The degree of vagueness v_{cc_a} greedily weights the edge count from c to c_a. Having a domain of $0 \le v_{cc_a} < 1$, v_{cc_a} is inversely proportional to the number of senses in c that have the hypernym c_a. If $v_{cc_a} = 0$, then there is no ambiguity or vagueness to the concept, meaning that all senses of c generalize to the concept c_a and therefore, $d_{cc_a} = 0$ as there are no other abstract concepts to consider as closer abstracts. However, this would heavily bias concepts higher in the concept hierarchy, so only senses whose closest generalization is c_a are employed in calculating v_{cc_a}, i.e. if c_a is a superordinate of another common hypernym, the sense that generalizes to the subordinate hypernym first will not be counted in determining the degree of vagueness for the superordinate hypernym. Obviously, $v_{cc_a} \ne 1$ as this would indicate that no sense of c has the hypernym c_a.

The degree of abstraction for the common hypernym c_a is then calculated by taking the average of its two specializers' d values, $d_{c_1c_a}$ and $d_{c_2c_a}$:

$$c_a.d = \frac{d_{c_1c_a} + d_{c_2c_a}}{2} = \frac{v_{c_1c_a} \cdot Min[S_{c_1c_a}] + v_{c_2c_a} \cdot Min[S_{c_2c_a}]}{2} \tag{7}$$

We will consider $c_a.d$ as the determinant of the least abstract common hypernym of the concepts to be generalized.

4.2 The LACH Algorithm

The algorithm begins the generalization process by finding all common hypernyms amongst the original concepts in T_0. It will disregard common hypernyms that are superordinates of another common hypernym. For example two concepts may share a common path between both *Physical Entity* and *Entity*; *Entity* is a superordinate of *Physical Entity* so it will be initially disregarded. If the terms were generalized to *Physical Entity* they may then be rolled up the concept tree into *Entity*. However, this does not exclude *Entity* as a possible common hypernym on another level of the tree, T_k where $k > 0$.

Upon finding the common hypernyms, the algorithm inserts information about them into a queue A. Each element, *a*, in the queue maintains specific information about one common hypernym. It is denoted by $a = \{d, \ell, c_1, c_2, c_a\}$ where: *a.d* is degree of abstraction of the stored hypernym, $a.\ell$ is the next abstraction level (where $a.\ell = \ell + 1$ and ℓ is the current level of abstraction), $a.c_1$ and $a.c_2$ are the common hypernym's specializers, and $a.c_a$ is the common hypernym itself. All elements a in queue A are ordered by the degree of abstraction, $a.d$, in ascending order.

The algorithm removes an element a from the front of the queue and inserts its common abstract, $a.c_a$ into the level of the concept hierarchy given by $a.\ell$. If $a.c_a$ is inserted into a new level where $T_{a.\ell}$ is empty, the original concepts that are not generalized by $a.c_a$ are then copied into $T_{a.\ell}$ as well. $a.c_a$ is then compared to these concepts looking for new common hypernyms. Any new common hypernyms found are then inserted into A. However, if $a.c_a$ is inserted into a level that already exists (i.e. $T_{a.\ell}$ is not empty), the algorithm must check to see if the lower level concepts that are not specializers of $a.c_a$ already exist at $T_{a.\ell}$. If they do not, it will copy them into $T_{a.\ell}$. It will then compare $a.c_a$ with the other concepts at $T_{a.\ell}$ that have differing lower level concepts as specializers from that of $a.c_a$.

This process will be continued until either (1) a generalization threshold is met (e.g. the size of the dataset has been reduced below a given size), (2) a specified success condition is met (e.g. the first common hypernym that encompasses all of the original terms is found), or (3) until the queue, A, is empty (i.e. no more common hypernyms have been found). The prototype currently assumes that it will run until A is empty, but the other mentioned conditions could be specified in the algorithm. When one of the stopping conditions is met the algorithm returns the fuzzy concept hierarchy, T, providing the common hypernyms for the original terms.

4.3 Example Execution of LACH

We now overview a simple example reflecting the execution of the LACH algorithm; how it finds the least abstract common hypernym and generates a fuzzy concept hierarchy. The example will take as input the concepts *cherry*, *C*, burgundy, *B*, and *lemon*, *L*, as presented in Figure 14. We explain in some detail the first few steps of the execution and then highlight significant cases from the subsequent steps.

The algorithm initially creates the first level of the tree T_0 inserting *C*, *B*, and *L*. Then their common hypernyms are found and inserted into the queue A. So initially the queue contains the entries: *(1.042, 1, B, C, R for 'red, redness'), (1.175, 1, L, C, FT for 'fruit tree'), (1.175, 1, L, C, EF for 'edible fruit'), (1.550, 1, L, C, Ch.C for 'chromatic color'), (1.800, 1, L, B, Ch.C), (2.667, 1, B, C, S for 'substance, matter')* and *(4.067, 1, L, B, S)*.

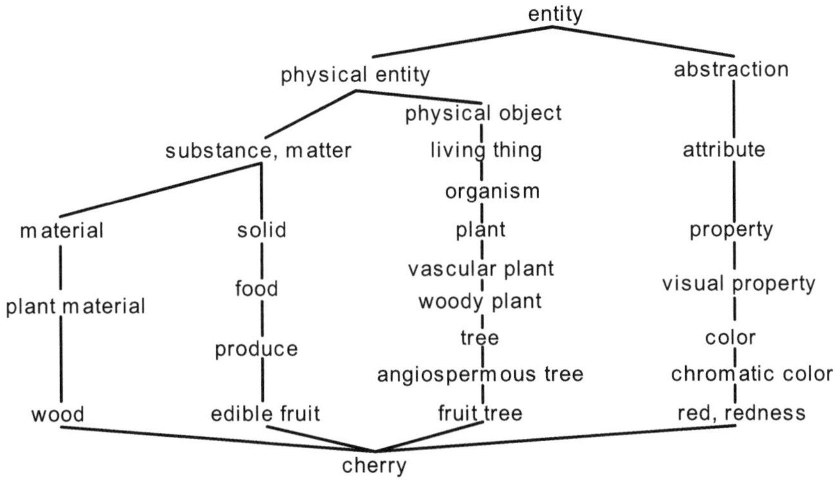

Fig. 13 Set of hypernym chains for CHERRY presented in the form of a Fuzzy Concept Hierarchy

Let us explain the formation of the first entry in some detail and the others follow similarly. The last 4 elements in the entry mean that *cherry* and *burgundy* have *redness* as a common generalizer at the next level, 1 (i.e. T_1), of the tree being constructed. To compute the degree of abstraction *R.d*, we first note that we are using 4 senses for *cherry* (see Fig. 8) and 3 senses for *burgundy* (see Fig. 12). From equation 5 we then have for *B*: $v_{BR} = 1 - \frac{1}{3} = \frac{2}{3}$ and for *C*: $v_{CR} = 1 - \frac{1}{4} = \frac{3}{4}$. The shortest generalization path for *B* to *R* is 2 (see Fig. 3) and for *C* to *R* is 1 (see Fig. 13) so from equation 6 we obtain $d_{BR} = \frac{2}{3} \cdot 2 = 1.33$ and $d_{CR} = \frac{3}{4} \cdot 1 = 0.75$. Finally using equation 7 we see that *R.d* is the average of these two values, $R.d = \frac{1.33 + 0.75}{2} = 1.042$.

The first entry in the front of the queue, representing generalization of *B* and *C*, is removed and inserted into the next level T_1 of the tree. As a result of this step new combinations can arise and since *R* and *L* have the common abstract of chromatic color, *Ch.C*, the entry *(0.800, 2, R, L, Ch.C)* is formed and inserted into A. Next a cleanup function removes two queue entries, *(1.550, 1, L, C, Ch.C)* and *(1.800, 1, L, B, Ch.C)*, because they are subsumed by this new entry *(0.800, 2, R, L, Ch.C)*. If they were not removed they would just add another iteration to the loop with no effect on the final result.

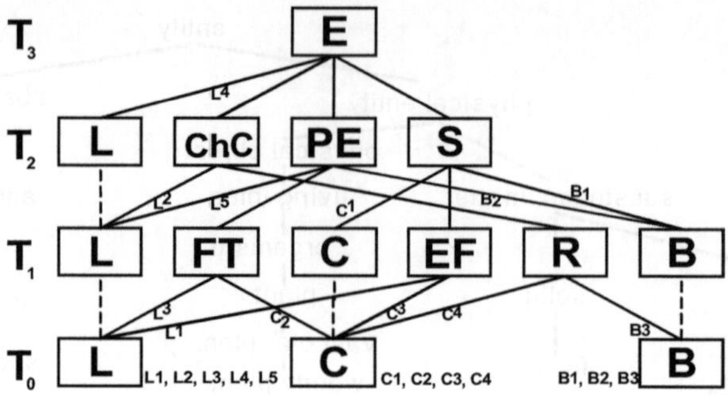

Fig. 14 Fuzzy concept hierarchy for the terms *burgundy*, *cherry* and *lemon* generated by the LACH algorithm

Continuing in this fashion *Ch.C* is inserted into T_2 but all original concepts are represented by *Ch.C*, so no terms are copied into the new level. The next entry from the queue is then inserted, *FT* into T_1, and a new combination *(0.833, 2, FT, B, PE for 'physical entity')* is found and put into A. Using this entry, *PE* is inserted into T_2, but we do not copy *L* into T_2 as *L* is an indirect specializer of *PE* (i.e. *PE* is generalization of *FT*, which is an abstract of *L*). Another new combination *(0.000, 3, PE, Ch.C, E for 'entity')* is then found and inserted at the front of A. This process continues until S is inserted into T_2 and no new combinations are found. A is now empty and a function that checks for consistent representation of original senses finds that *cherry* is not fully represented on T_1 so it copies the sense of *cherry, wood* to T_1 (sense *C1*, marked as dotted line in Figure 14). Now two new common hypernyms are formed, *(2.000, 2, C, EF, S)*, *(2.667, 2, B, C, S)* and inserted into A. After several more steps the first entry in the front of the queue is *(0.000, 3, L, Ch.C, E)* and so *E* is inserted into T_3. Since all of the original concepts are represented by *E*, no terms are copied into the new level. Then the entry *(0.000, 3, L, S, E)* is used to add generalization of *S* to *E* (the link *L* to *E* already exists, since it was entered when *(0.000, 3, L, Ch.C, E)* was processed). The cleanup function removes subsumed entry *(0.000, 3, L, PE, E)* from A and so again A is empty. The membership checking function is then invoked and as all senses are fully represented in all the levels of T there is no change. So finally the queue is still empty and the algorithm terminates with the completed form of the tree as seen in Figure 14. In the figure the numbers denote the sense numbers being contained (or propagated). Solid lines depict a generalization, dashed lines depict a temporary specialization.

Every original concept is represented by a unity at any level in the tree (as defined in equations 2 and 3). A concept c fully belongs to its abstract c_a if all senses represented by c are characterized by concept c_a, otherwise the vote representing multiple senses of c may be spread across multiple hypernyms (as shown in Table 1). For example, the original concept *c* at level T_2 is represented by the fuzzy set $\{Ch.C|¼$ (for sense *c*4 *red, redness*); *PE*|¼ (for sense *c*2 *fruit tree*); *S*|½ (for senses *c*1 *cherry, word* and *c*3 *edible fruit*)$\}$.

Table 1. Propagation of membership values generated for FCH in Figure 14

	L	C	B
Level 3 (T_3)			
E	1.0	1.0	1.0
Level 2 (T_2)			
L	0.2	0.0	0.0
Ch. C	0.2	0.25	0.33
PE	0.4	0.25	0.33
S	0.2	0.5	0.33
Level 1 (T_1)			
L	0.6	0.0	0.0
FT	0.2	0.25	0.0
C	0.0	0.25	0.0
EF	0.2	0.25	0.0
R	0.0	0.25	0.33
B	0.0	0.0	0.66
Level 0 (T_0)			
L	1.0	0.0	0.0
C	0.0	1.0	0.0
B	0.0	0.0	1.0

Given the input of *cherry*, *burgundy* and *lemon* the result of the algorithm is a fuzzy concept hierarchy constructed with common hypernyms depicted in Figure 14. The first possible common hypernym for all of the concepts in T_0 occurs in no later than the $(n-1)^{th}$ level where n is the number of concepts in T_0. However, common hypernyms for all of the concepts in T_0 may also occur in the nth level or above. These concepts will not be as close abstractly to the original concepts, but they may provide a more accurate generalization as they may cover multiple senses of some of the original concepts. Recall that at each level of the fuzzy concept hierarchy each concept and all of its senses are fully represented.

The common hypernym that encompasses each of the concepts at T_0 on the lowest level, on or before the $(n-1)^{th}$ level, and with the lowest d, is the least abstract common hypernym for the concepts at T_0. Given some background knowledge that was not used to influence the algorithm, this may not be the preferred common hypernym; so the fuzzy concept hierarchy is returned for cases where a further choice could be made. Also, by having this returned fuzzy concept hierarchy, the result may be rolled up or drilled down to get closer to a specific generalization threshold for the data set.

5 Conclusions and Future Research

The algorithm presented in this paper will return a fuzzy concept hierarchy, T, of common hypernyms for a set of given concepts. The results returned are the best choices for generalization without any background knowledge. It does not take into account any background knowledge, though it may be modified to allow human assessment to influence the generated results in the future. Our approach allows the unsupervised automated construction of fuzzy concept hierarchies. The path between the concepts being generalized and each level in the fuzzy concept hierarchy generalizes the concepts to their least abstract common hypernyms.

As with any unsupervised automated generalization, this algorithm may not provide a perfect solution. However, in our opinion given an appropriate set of terms to be generalized, an accurate least abstract common hypernym and a fuzzy concept hierarchy are achievable. Given the common complaint that extended expert knowledge is necessary for creation of individual-purpose ontologies, we believe the algorithm provides an intriguing opportunity for researchers interested in unsupervised automated summarization of descriptive values.

The problem of appropriate sense generalization can be solved based on multiple approaches. We can for instance present the whole concept hierarchy to the user, allowing them to decide which level is the most appropriate or convenient for her purposes. Obviously she should not be forced to choose only one level; she may want to pick up terms coming from different levels of abstraction. Another way to decide is to choose the approximate number of abstract concepts to be used to generalize the inserted terms. In such a case we may traverse the generated concept hierarchy starting from its top until we find a level containing the number of concepts responding to our preferences.

The algorithm is not limited to WordNet, but it is limited to the same hypernym/sense structure that WordNet uses to build its concept hierarchies. Given a customized set of ontologies (beyond WordNet), the algorithm could provide accurate unsupervised generalizations. A customized concept hierarchy could then be implemented for any data set, and this algorithm would provide effective results. We have applied this approach to the data mining approach of generalization but expect to examine the use of hierarchies developed in this manner for extraction of association rules.

Further work is needed to ascertain if there are more accurate measures of determining the least abstract common hypernym. One prospect is to take a lead from WordNet, allowing a polysemy count of a hypernym to influence its value as a candidate for the least abstract common hypernym. WordNet recognizes that polysemy is a factor in familiarity of words; so given a choice between two concepts with very small degrees of abstraction, choosing a hypernym that has a higher polysemy (multiple meanings) count, i.e. the more familiar term, may provide a more accurate result. Polysemy is already involved in determining the degree of abstraction for each concept and could play a much greater role in determining the least abstract common hypernym in future revisions of the algorithm.

In its first version WordNet maintained a semantic concordance that measured the relative frequency of a given word's senses. Unfortunately, the current versions of the WordNet no longer support this method so we were not able to use the frequency number to our advantage, i.e. to allow for a more accurate estimate of what sense is more prominent and then to achieve a higher accuracy rate at building a blind concept hierarchy. This element however could be easily emphasized when working with sources providing such capability. Synonyms, antonyms and meronyms may be of some use in helping identify specific senses when automating the concept hierarchy construction. Troponyms and entailment will probably be of less use to the concept hierarchies as they are properties of verbs which will are uncommon in the generalization steps of attribute oriented induction.

References

1. WordNet is an on-line lexical reference system, developed by Princeton University Cognitive Science Laboratory and available at: http://wordnet.princeton.edu/. For this work we used version 2.1.
2. Carter CL & Hamilton HJ, "Efficient Attribute-Oriented Generalization for Knowledge Discovery from Large Databases", *IEEE Transactions on Knowledge and Data Engineering, 10*(2), 1998, pp. 193-208.
3. Cai Y, Cercone N, & Han J, "Attribute-Oriented Induction in Relational Databases", *Proc. IJCAI-89 Workshop on Knowledge Discovery in Databases*, Detroit, MI, USA, 1989, pp. 26 –36.
4. Cai Y, Cercone N, & Han J, "Knowledge discovery in databases: An attribute-oriented approach", *Proc. 18th Int. Conf. Very Large Data Bases*, Vancouver, Canada, 1992, pp. 547-559.
5. Han J & Fu Y, "Discovery of multiple-level association rules from large databases", *Proc. 21st Int. Conf. Very Large Data Bases*, Zurich, Switzerland, 1995, pp. 420-431.
6. Cai Y, Cercone N, & Han J, "Data-Driven Discovery of Quantitative Rules in Relational Databases," *IEEE Transactions on Knowledge and Data Engineering, 5*(1), 1993, pp. 29-40.
7. Han J, "Towards Efficient Induction Mechanisms in Database Systems", *Theoretical Computing Science, 133*, 1994, pp. 361-385.
8. Han J, Kawano H, Nishio S & Wang W, „Generalization-based data mining in object-oriented databases using an object cube model", *Data & Knowledge Engineering, 25*, 1998, pp. 55-97.

9. Han J & Kamber M, *Data Mining: Concepts and Techniques*, Morgan Kaufmann, New York, NY, 2000.
10. Hamilton HJ, Cercone N & Hilderman RJ, "Attribute-oriented induction using domain generalization graphs", *Proc. 8th IEEE Int'l Conf. on Tools with Artificial Intelligence*, Toulouse, France, 1996, pp. 246-253.
11. Hamilton HJ, Cercone N & Hilderman RJ, "Data mining in large databases using domain generalization graphs", *Journal of Intelligent Information Systems, 13*(3), 1999, pp. 195-234.
12. Hwang HY & Fu WC, "Efficient Algorithms for Attribute-Oriented Induction", *Proceeding of The First International Conference on Knowledge Discovery and Data Mining*, Montreal, Canada, August 1995, pp. 168-173.
13. Han J, Fu Y, & Tang S, "Advances of the DBLearn System for Knowledge Discovery in Large Databases," Proc. IJCAI, Int'l Joint Conf. Artificial Intelligence, Montreal, August 1995, pp. 2049-2050.
14. Lee DH & Kim MH, "Database summarization using fuzzy ISA hierarchies", *IEEE Transactions on Systems, Man, and Cybernetics - part B, 27*(1), 1997, pp. 68-78.
15. Lee KM, "Mining generalized fuzzy quantitative association rules with fuzzy generalization hierarchies", *Proc. Joint 9th IFSA World Congress and 20th NAFIPS Int'l Conf.*, Vancouver, Canada, 2001, pp. 2977-2982.
16. Cubero JC, Medina JM, Pons O & Vila MA, "Data Summarization in Relational Databases Through Fuzzy Dependencies", *Information Sciences, 121*(3-4), 1999, pp. 233-270.
17. Angryk R, Ladner R & Petry F "Mining Generalized Knowledge from Imperfect Data," Proceedings of the 10th International Conference on Information Processing and Management of Uncertainty in Knowledge-Based Systems (IPMU '04), Perugia, Italy, July 2004, pp. 739-746.
18. Barbu C, Angryk R, Petry F, Simina M, "Information Filtering via Fuzzy Hierarchical Induction, " Proceedings of the IEEE International Conference on Systems, Man & Cybernetics (SMC '04), Hague, Netherlands, October 2004, pp. 3576-3581.
19. Angryk R, Petry F, "Discovery of Abstract Knowledge from Fuzzy Relational Databases," *Proceedings of the EUROFUSE Workshop on Data and Knowledge Engineering (EUROFUSE '04)*, Warsaw, Poland, September 2004, pp. 23-32.
20. Angryk R, Petry F, "Discovery of generalized knowledge from Proximity- and Similarity-based Fuzzy Relational Databases," *Special Topic Issue on Advances in Fuzzy Database Technology in International Journal of Intelligent Systems*, to appear.
21. Miller GA, Beckwith R, Fellbaum C, Gross D, Miller K, Introduction to WordNet: an on-line lexical database, Internat. Journal of Lexicography 3(4) (1990) 235-244.
22. Miller GA, 1990, "Nouns in WordNet: A Lexical Inheritance System", International Journal of Lexicography, Vol. 3, No. 4, 245-264.
23. Miller GA, "WordNet: A Lexical Database for English", Communications of the ACM, November 1995, 38(11), pp. 39-41.
24. Miller GA, et. al., "Five Papers on WordNet", CSL Report 43, Cognitive Science Lab, Princeton University, 1990. ftp://ftp.cogsci.princeton.edu/pub/wordnet/5papers.pdf
25. WordNet: An Electronic Lexical Database (Language, Speech, and Communication) by Christiane Fellbaum (Editor), MIT Press, 1998.
26. Calmet J, Daemi A, "From Entropy to Ontology", *Cybernetics and systems 2004 - ATFrom Agent Theory to Agent Implementation: 4*, R. Trappl, Vol. 2, 2004, pp. 547–551.
27. Calmet J., Daemi A, "From Ontologies to Trust through Entropy", *Proc. Int'l Conf. on Advances in Intelligent Systems - Theory and Applications*, Luxembourg, November, 2004.

28. Resnik P, "Using Information Content to Evaluate Semantic Similarity in a Taxonomy", *Proc. 14th Int'l Joint Conf. on Artificial Intelligence*, 1995, pp. 448-453.
29. Budanitsky A, Hirst G, "Semantic Distance in WordNet: An Experimental Application-Oriented Evaluation of Five Measures", *Proc. Workshop on WordNet and Other Lexical Resources, in the NAACL,* Pittsburgh, PA, USA, 2001.

APPENDIX: The Least Abstract Common Hypernym (LACH) Algorithm

LACH $(c_1, c_2, ..., c_n)$

Input: $c_1, c_2, ..., c_n$ - concepts to be generalized

1. Set $\ell = 0$
2. Insert $c_1, c_2, ..., c_n$ into T_ℓ
3. Find all common hypernyms for all concepts in T_ℓ, each represented as a
4. Measure each common hypernym's degree of abstraction $a.d$
5. Assign each common hypernym's level $a.\ell = \ell + 1$
6. Insert found common hypernyms in queue A ordered by their degree of abstraction d s.t. $a_1, a_2, \ldots, a_n$ and $a_{i-1}.d \le a_i.d$ for $1 < i \le n$
7. Until A is empty:
8. Dequeue a, the first value, from A
9. Insert a into T_ℓ where $\ell = a.\ell$
10. Copy concepts that are not represented by a into T_ℓ
11. Find all common hypernyms of a and all concepts in T_ℓ, each represented as b
12. Measure every b's degree of abstraction $b.d$
13. Assign every b's level $b.\ell = \ell + 1$
14. Insert every b into A

Steps 15 – 26 ensure that every concept from T_0 is fully represented on each level of the concept hierarchy. This is completed on a step-by-step process so that as each level in T fully represents the original concepts, new common hypernyms can be found for the next level in the hierarchy.

15. If A is empty
16. For each level ℓ in T where $1 \le \ell \le T.height()$
17. If (all concepts at T_0 are not fully represented at T_ℓ) then
18. Copy senses of underrepresented concepts into T_ℓ
19. Find all common hypernyms for all concepts in T_ℓ, each represented as a
20. Measure every a's degree of abstraction $a.d$
21. Assign every a's level $a.\ell = \ell + 1$
22. Insert every a into A
23. Exit Loop (return to step 7)
24. End If (step 17)
25. End Loop (step 16)
26. End If (step 15)
27. End Loop (step 7)
28. Return T

Beyond the Semantic Web: Fuzzy Logic-Based Web Intelligence

Masoud Nikravesh
BISC Program, EECS Department and
Imaging and Informatics- Life Sciences Division, Lawrence Berkeley National Lab
University of California, Berkeley, CA 94720, USA
nikravesh@cs.berkeley.edu,
http://www-bisc.cs.berkeley.edu
Tel: (510) 643-4522, Fax: (510) 642-5775

Abstract: *World Wide Web search engines have become the most heavily-used online services, with millions of searches performed each day. Their popularity is due, in part, to their ease of use. It is important to note that while the Semantic Web is dissimilar in many ways from the World Wide Web, the* ***Semantic Web*** *provides a common framework that allows* ***data*** *to be shared and reused across application, enterprise, and community boundaries through the World Wide Web. In this paper, we would like to go beyond the traditional semantic web which has been defined mostly as a mesh or distributed databases within the World Wide Web. For this reason, our view is that "Before one can use the power of semantic web, the relevant information has to be mined through the search mechanism and logical reasoning". Therefore, a right search engine model that can be integrated into a web-based distributed database system capable of reasoning is a need to enhance the power of the current semantic web systems. The central tasks for the most of the search engines can be summarize as 1) query or user information request- do what I mean and not what I say!, 2) model for the Internet, Web representation-web page collection, documents, text, images, music, etc, and 3) ranking or matching function-degree of relevance, recall, precision, similarity, etc. Design of any new intelligent search engine should be at least based on two main motivations: 1) The web environment is, for the most part,*

M. Nikravesh: *Beyond the Semantic Web: Fuzzy Logic-Based Web Intelligence*, StudFuzz **204** 149–242 (2006)
www.springerlink.com

unstructured and imprecise. To deal with information in the web environment what is needed is a logic that supports modes of reasoning which are approximate rather than exact. While searches may retrieve thousands of hits, finding decision-relevant and query-relevant information in an imprecise environment is a challenging problem, which has to be addressed and 2) Another, and less obvious, is deduction in an unstructured and imprecise environment given the huge stream of complex information. In this paper, we will first present the state of the search engines and internet. Then we will focus on development of a framework for reasoning and deduction in the web. A web-based model to decision model for analysis of structured database will be presented. A framework to incorporate the information from web sites into the search engine will be presented as a model that will go beyond current semantic web idea.

1. Introduction

What is Semantic Web? "The Semantic Web is an extension of the current web in which information is given well-defined meaning, better enabling computers and people to work in cooperation." -- *Tim Berners-Lee, James Hendler, Ora Lassila, The Semantic Web, Scientific American, May 2001*

"The **Semantic Web** provides a common framework that allows **data** to be shared and reused across application, enterprise, and community boundaries. It is a collaborative effort led by W3C with participation from a large number of researchers and industrial partners. It is based on the Resource Description Framework (RDF), which integrates a variety of applications using XML for syntax and URIs for naming." – W3C organization (http://www.w3.org/2001/sw/)

"Facilities to put machine-understandable data on the Web are becoming a high priority for many communities. The Web can reach its full potential only if it becomes a place where data can be shared and processed by automated tools as well as by people. For the Web to scale, tomorrow's programs must be able to share and process data even when these programs have been designed totally independently. The Semantic Web is a vision: the idea of having data on the web defined and linked in a way that it can be used by machines not just for display purposes, but for automation, integration and reuse of data across various applications." (http://www.w3.org/2001/sw/).

Semantic Web is a mesh or network of information that are linked up in such a way that can be accessible and be processed by machines easily, on a global scale. One can think of Semantic Web as being an efficient way of representing and sharing data on the World Wide Web, or as a globally linked database. It is important to mention that Semantic Web technologies are still very much in their infancies and there seems to be little consensus about the likely characteristics of such system. It is also important to keep in mind that the data that is generally hidden in one way or other is often useful in some contexts, but not in others. It is also difficult to use on a large scale such information, because there is no global system for publishing data in such a way as it can be easily processed by anyone. For example, one can think of information about local hotels, sports events, car or home sales info, insurance data, weather information stock market data, subway or plane times, Major League Baseball or Football statistics, and television guides, etc..... All these information are presented by numerous web sites in HTML format. Therefore, it is difficult to use such data/information in a way that one might wanted to do so.

To build any semantic web-based system, it will become necessary to construct a powerful logical language for making inferences and reasoning such that the system to become expressive enough to help users in a wide range of situations. This paper will try to develop a framework to address this issue. Figure 1 shows Concept-Based Web-Based Databases for Intelligent Decision Analysis; a framework for next generation of Semantic Web. In this paper, we will first present the state of the search engines and internet. Then we will focus on development of a framework for reasoning and deduction in the web. A web-based model to decision model for analysis of structured database will be presented. A framework to incorporate the information from web sites into the search engine will be presented as a model that will go beyond current semantic web idea (Figure 1).

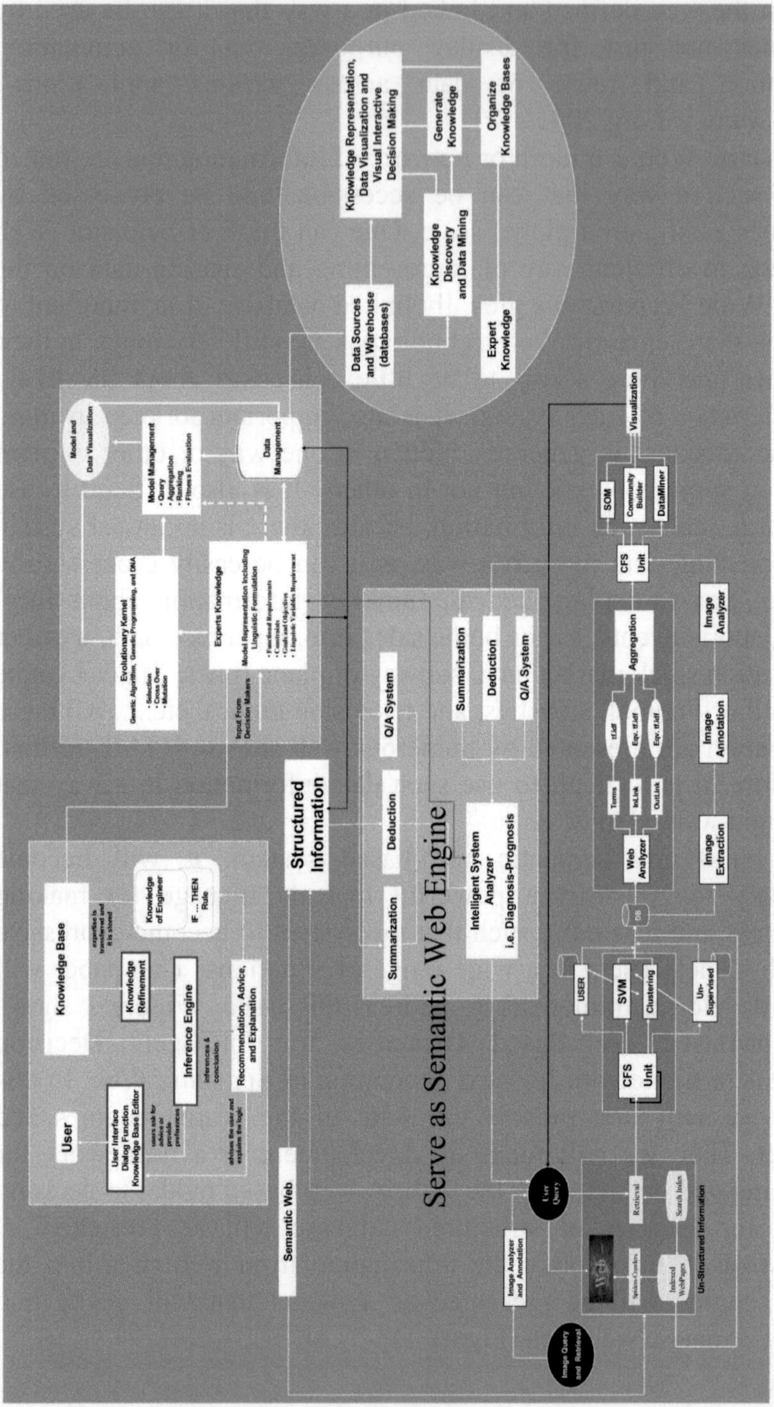

Figure 1. Concept-Based *Web-Based Databases for* Intelligent Decision Analysis; *a framework for next generation of Semantic Web*

2. Search Engine and the Internet

In this section, we present the state search engines and the Internet. Under leadership of DARPA, ARPANET has been designed through close collaboration with UCLA during 1962-1969, 1970-1973, and 1974-1981. Initially designed to keep military sites in communication across the US. In 1969, ARPANET connected researchers from Stanford University, UCLA, UC Santa Barbara and the University of Utah. The Internet community formed in 1972 and the Email is started in 1977. While initially a technology designed primarily for needs of the U.S. military, the Internet grew to serve the academic and research communities. More recently, there has been tremendous expansion of the network both internationally and into the commercial user domain. ***Table 1*** shows the evolution of the Internet, World Wide Web, and Search Engines.

There are many publicly available Web search engines, but users are not necessarily satisfied with speed of retrieval (i.e., slow access) and quality of retrieved information (i.e., inability to find relevant information). It is important to remember that problems related to speed and access time may not be resolved by considering Web information access and retrieval as an isolated scientific problem. An August 1998 survey by Alexa Internet (<alexa.com>) indicates that 90% of all Web traffic is spread over 100,000 different hosts, with 50% of all Web traffic headed towards the top 900 most popular sites. Effective means of managing uneven concentration of information packets on the Internet will be needed in addition to the development of fast access and retrieval algorithms (Kabayashi and Takeda 2000).

During 80, most of the advances of the automatic document categorization and IR were based on knowledge engineering. The models were built manually using expert systems capable of taking decision. Such expert system has been typically built based on a set of manually defined rules. However, the bottleneck for such manual expert systems was the knowledge acquisition very similar to expert system. Mainly, rules needed to be defined manually by expert and were static. Therefore, once the database has been changed or updated the model must intervene again or work has to be repeated

Table 1. Understanding and History of Internet, World Wide Web and Search Engine;

Search Engine and Internet	Date	Developer	Affiliation	Comments
ARPANET	1962 –1969 1970-1973 1974-1981	UCLA	Under Leadership of DARPA	Initially designed to keep military sites in communication across the US. In 1969, ARPANET connected researchers from Stanford University, UCLA, UC Santa Barbara and the University of Utah. Internet community formed (1972). Email started (1977).
ALOHANET	1970		University of Hawaii	
USENET	1979	Tom Truscott & Jim Ellis Steve Bellovin	Duke University & University of North Carolina	The first newsgroup.
ARPANET	1982-1987	Bob Kahn & Vint Cerf	DARPA & Stanford University	ARPANET became “Internet”. Vinton Cerf “Father of the Internet”. Email and Newsgroups used by many universities.
CERT	1988-1990	Computer Emergency Response Team		Internet tool for communication. Privacy and Security. Digital world formed. Internet worms & hackers. The World Wide Web is born.
Archie through FTP	1990	Alan Ematage	McGill University	Originally for access to files given exact address. Finally for searching the archive sites on FTP server, deposit and retrieve files.
Gopher	1991	A team led by Mark MaCahill	University of Minnesota	Gopher used to organize all kinds of information stored on universities servers, libraries, non-classified government sites, etc. Archie and Veronica, helped Gopher (Search utilities).
World Wide Web "alt.hypertext	1991	Tim Berners-Lee	CERN in Switzerland	The first World Wide Web computer code. "alt.hypertext." newsgroup with the ability to combine words, pictures, and sounds on Web pages
Hyper Text Transfer Protocol (HTTP).	1991	Tim Berners-Lee	CERN in Switzerland	The 1990s marked the beginning of World Wide Web which in turn relies on HTML and Hyper HTTP. Conceived in 1989 at the CERN Physics Laboratory in Geneva. The first demonstration December 1990. On May 17, 1991, the World Wide Web was officially started, by granting HTTP access to a number of central CERN computers. Browser software became available-Microsoft Windows and Apple

				Macintosh
	1992			The first audio and video broadcasts:"MBONE." More than 1,000,000 hosts.
Veronica	1993	System Computing Services Group	University of Nevada	The search Device was similar to Archie but search Gopher servers for Text Files
Mosaic	1993	Marc Andeerssen	NCSA (the National Center for Supercomputin g Applications); University of Illinois at Urbana Champaign	Mosaic, Graphical browser for the World Wide Web, were developed for the Xwindows/UNIX, Mac and Windows.
World Wide Web Wanderer; the first Spider robot	1993	Matthew Gary	MIT	Developed to count the web servers. Modified to capture URLs. First searchable Web database, the Wandex.
ALIWEB	1993	Martiijn Koster	Now with Excite	Archie-Like Indexing of the Web. The first META tag
JumpStation, World Wide Web Worm .	1993		NASA	Jump Station developed to gathere document titles and headings. Index the information by searching database and matching keywords. WWW worm index title tags and URLs.
Repository-Based Software Engineering (RBSE) Spider	1993		NASA	The first relevancy algorithm in search results, based on keyword frequency in the document. Robot-Driven Search Engine Spidered by content.
	1994			Broadcast over the M-Bone. Japan's Prime Minister goes online at www.kantei.go.jp. Backbone traffic exceeds 10 trillion bytes per month.
Netscape and Microsoft's Internet	1994-1998	Microsoft and Netscape	Microsoft and Netscape	Added a user-friendly point-and-click interface for browsing

Explorer				
Netscape	1994	Dr. James H. Clark and Marc Andreessen		The company was founded in April 1994 by Dr. James H. Clark, founder of Silicon Graphics, Inc. and Marc Andreessen, creator of the NCSA Mosaic research prototype for the Internet. June 5, 1995 - change the character of the World Wide Web from static pages to dynamic, interactive multimedia.
Galaxy	1994	Administered by Microelectronics and computer Technology Corporation	Funded by DARPA and consortium of technologies companies and original prototype by MADE program.	Provided large-scale support for electronic commerce and links documents into hierarchical categories with subcategories. Galaxy merged into Fox/News in 1999.
WebCrawler	1994	Brian Pinkerton	University of Washington	Search text of the sites and used for finding information in the Web. AOL purchased WebCrawler in 1995. Excite purchased WebCrawler in 1996.
Yahoo!	1994	David Filo and Jerry Yang	Stanford University	Organized the data into searchable directory based on simple database search engine. With the addition of the Google, Yahoo! Is the top-referring site for searches on the Web. It led also the future of the internet by changing the focus from search retrieval methods to clearly match the user's intent with the database.
Lycous	1994	Michael Mauldin	Carnegie Mellon University	New features such as ranked relevance retrieval, prefix matching, and word proximity matching. Until June 2000, it had used Inktomi as its back-end database provide. Currently, FAST a Norwegian search provider, replaced the Inktomi.
Excite	1995	Mark Van Haren, Ryan McIntyre, Ben Lutch, Joe Kraus, Graham Spencer, and Martin Reinfried	Architext Sofware	Combined search and retrieval with automatic hypertext linking to document and includes subject grouping and automatic abstract algorithm. IT can electronically parse and abstract from the web.

Infoseek	1995	Steve Kirsch (now with Propel)	Infoseek	Infoseek combined many functional elements seen in other search tools such as Yahoo! And Lycos, but it boasted a solid user-friendly interface and consumer-focused features such as news. Also speed in which indexed Web sites and then added them to its live search database.
AltaVista	1995	Louis Monier, with Mike Burrows	Digital Equipment Corporation	Speed and the first "Natural Language" queries and Boolean operators. It also proved a user-friendly interface and the first search engine to add a link to helpful search tips below search field to assist novice searchers.
MetaCrawler	1995	Erick Selberg and Oren Etizinoi	University of Washington	The first Meta search engine. Search several search engines and reformat the results into a single page.
SavvySearch	1995	Daniel Dreilinger	Colorado State University	Meta Search which was included 20 search engines. Today, it includes 200 search engine.
Inktomi and HotBot	1994-1996	Eric Brewer and Paul Gauthier	University of California-Berkeley Funded by ARPA	Cluster inexpensive workstation computers to achieve the same computing power as expensive super computer. Powerful search technologies that made use of the clustering of workstations to achieve scaleable and flexible information retrieval system. HotBot, powered by Inktomi and was able to rapidly index and spider the Web and developing a very large database within a very short time.
LookSmart	1996	Mr Evan Thornley	LookSmart	Delivers a set of categorized listing presented in a user-friendly format and providing search infrastructure for vertical portals and ISPs.
AskJeeves	1997	Davis Warthen and Garrett Gruener	AskJeeves	It is built based on a large knowledge base on pre-searched Web sites. It used sophisticated, natural-language semantic and syntactic processing to understand the meaning of the user's question and match it to a 'question template" in the knowledge base.

GoTo	1997	Bill Gross	Indealab!	Auctioning off search engine positions. Advertisers to attach a value to their search engine placement.
Snap	1997	Halsey Minor, CNET Founder	CNET, Computer Network	Redefining the search engine space with a new business model; "portal" as first partnership between a traditional media company and an Internet portal.
Google	1997-1998	Larry Page and Sergey Brin	Stanford University	PageRank™ to deliver highly relevant search results based on proximity match and link popularity algorithms. Google represent the next generation of search engines.
Northern Light	1997	Team of librarians, software engineers, and information industry	Northern Light	To Index and classify human knowledge and has two database 1) contains an index to the full text of millions of Web pages and 2) includes full-text articles from a variety of sources. It searches both Web pages and full-text articles and sorts its search results into folders based on keywords, source, and other criteria.
AOL, MSN and Netscape	1998	AOL, MSN and Netscape	AOL, MSN and Netscape	Search service for the users of services and software
Open Directory	1998	Rick Skrenta and Bob Truel	dmoz	Open directory
Direct Hit	1998	Mike Cassidy	MIT	Direct Hit is dedicated to providing highly relevant Internet search results. Direct Hit's highly scalable search system leverages the searching activity of millions of Internet searchers to provide dramatically superior search results. By analyzing previous Internet search activity, Direct Hit determines the most relevant sites for your search request.
FAST Search	1999	Isaac Elsevier	FAST; Norwegian Company- All the Web	High-capacity search and real-time content matching engines based on the All the Web technology. Using Spider technology to index pages very rapidly. FAST can index both Audio and Video files.

a new if the system to be ported to a completely different domain. By explosion of the Internet, these bottlenecks are more obvious today. During 90, new direction has been merged based on machine learning approach. The advantage of this new approach is evident compared to the previous approach during 80. In machine learning approach, most of the engineering efforts goes towards the construction of the system and mostly is independent of the domain. Therefore, it is much easier to port the system into a new domain. Once the system or model is ported into a new domain, all that is needed is the inductive, and updating of the system from a different set of new dataset, with no required intervention of the domain expert or the knowledge engineer. In term of the effectiveness, IR techniques based on machine learning techniques achieved impressive level of the performance and for example made it possible automatic document classification, categorization, and filtering and making these processes viable alternative to manual and expert system models.

During the recent years, applications of fuzzy logic and the Internet from Web data mining to intelligent search engine and agents for Internet applications have greatly increased (Nikravesh, 2002; Nikravesh et al., 2002, 2003a, 2003b, 2003c; Nikravesh and Choi 2003, Loia et al. 2002, 2003; Nikravesh and Azvine, 2001, 2002; Takagi et al., 2002a, 2002b). Martin (2001) concluded that semantic web includes many aspects, which require fuzzy knowledge representation and reasoning. This includes the fuzzification and matching of concepts. In addition, it is concluded that fuzzy logic can be used in making useful, human-understandable, deduction from semi-structured information available in the web. It is also presented issues related to knowledge representation focusing on the process of fuzzy matching within graph structure. This includes knowledge representation based on conceptual graphs and Fril++. Baldwin and Morton (1985) studied the use of fuzzy logic in conceptual graph framework. Ho (1994) also used fuzzy conceptual graph to be implemented in the machine-learning framework. Baldwin (2001) presented the basic concept of fuzzy Bayesian Nets for user modeling, message filtering and data mining. For message filtering the protoype model representation has been used. Given a context, prototypes represent different types of people and can be modeled using fuzzy rules, fuzzy decision tree, fuzzy Bayesian Net or a fuzzy conceptual graph. In their study, fuzzy set has been used for better generalization. It has been also concluded that the new approach has many applications. For example, it can be used for personalization of web pages, intelligent filtering of the Emails, providing TV programs, books or movie and video of interest. Cao (2001) presented the fuzzy conceptual graphs for the semantic web. It is concluded that the use of conceptual graph and fuzzy logic is complementary for the semantic web. While conceptual graph provide a structure for natural language sentence, fuzzy logic provide a methodology for computing with words. It has been concluded that fuzzy conceptual graphs is suitable language for knowledge representation to be used by Semantic web. Takagi and Tajima (2001a, 2001b) presented the conceptual matching of text notes to be used by search engines. An new search engine proposed which conceptually matches keywords and the web pages. Conceptual fuzzy set has been used for context-dependent keyword expansion. A new structure for search engine has been proposed which can resolve the context-dependent word ambiguity using fuzzy conceptual matching technique. Berenji (2001) used Fuzzy Reinforcement Learning (FRL) for text data mining and Internet search engine. Choi (2001) presented a new technique, which integrates document index with perception index. The techniques can be used for refinement of fuzzy queries on the Internet. It has been concluded that the use of perception index in commercial search engine provides a framework to handle fuzzy terms (perception-based), which is further step toward a human-friendly, natural language-based interface for the Internet. Sanchez (2001) presented the concept of Internet-based fuzzy Telerobotic for the WWW. The system receives the information from human and has the capability for fuzzy reasoning. It has be proposed to use fuzzy applets such as fuzzy logic propositions in the form of fuzzy rules that can be used for smart data base search. Bautista and Kraft (2001) presented an approach to use fuzzy logic for user profiling in Web retrieval applications. The technique can be used to expand the queries and knowledge extraction related to a group of users with common interest. Fuzzy representation of terms based on linguistic qualifiers has been used for their study. In addition, fuzzy clustering of the user profiles can be used to construct fuzzy rules and inferences in order to modify queries. The result can be used for knowledge extraction from user profiles for marketing purposes. Yager (2001) introduced fuzzy aggregation methods for intelligent search. It is concluded that the new technique can increase the expressiveness in the queries. Widyantoro and Yen (2001) proposed the use of fuzzy ontology in search engines. Fuzzy ontology of term relations can be built automatically from a collection of documents. The proposed fuzzy ontology can be used for query refinement and to suggest narrower and broader terms suggestions during user search activity. Presser (2001) introduced fuzzy logic for rule-based personalization and can be implemented for personalization of newsletters. It is concluded that the use of fuzzy logic provide better flexibility and better interpretation which helps in keeping the knowledge bases easy to maintain. Zhang et al. (2001a) presented granular fuzzy technique for web search engine to increase Internet search speed and the Internet quality of service. The techniques can be used for personalized fuzzy web search engine, the personalized granular web search agent. While current fuzzy search engines uses keywords, the proposed technique provide a framework to not only use traditional fuzzy-key-word but also fuzzy-user-preference-based search algorithm. It is concluded that the proposed model reduces web search redundancy, increase web search relevancy, and decrease user's web search time. Zhang et al. (2001b) proposed fuzzy neural web agents based on granular neural network, which discovers fuzzy rules for stock prediction. Fuzzy logic can be used for web mining. Pal et al. (2002) presented issues related to web mining using soft computing framework. The main tasks of web mining based on fuzzy logic include information retrieval and generalization. Krisnapuram et al. (1999) used fuzzy c medoids and triimed medoids for clustering of web documents. Joshi and Krisnapuram (1998) used fuzzy clustering for web log data mining. Sharestani (2001) presented the use of fuzzy logic for network intruder detection. It is concluded that fuzzy logic can be used for approximate reasoning and handling detection of intruders through approximate matching; fuzzy rule and summarizing the audit log data. Serrano (2001) presented a web-based intelligent assistance. The model is an agent-based system which uses a knowledge-based model of the e-business that provide advise to user through intelligent reasoning and dialogue evolution. The main advantage of this system is based on the human-computer understanding and expression capabilities, which generate the right information in the right time.

3. NeuSearch™: Conceptual-Based Text Search and Question Answering System Using Structure and Unstructured Text Based on PNL, Neuroscience

In this section, we will focus on development of a framework for reasoning and deduction in the web.

World Wide Web search engines have become the most heavily-used online services, with millions of searches performed each day. Their popularity is due, in part, to their ease of use. The central tasks for the most of the search engines can be summarize as 1) query or user information request- do what I mean and not what I say!, 2) model for the Internet, Web representation-web page collection, documents, text, images, music, etc, and 3) ranking or matching function-degree of relevance, recall, precision, similarity, etc.
One can use clarification dialog, user profile, context, and ontology, into an integrated frame work to design a more intelligent search engine. The model will be used for intelligent information and knowledge retrieval through conceptual matching of text. The selected query doesn't need to match the decision criteria exactly, which gives the system a more human-like behavior. The model can also be used for constructing ontology or terms related to the context of search or query to resolve the ambiguity. The new model can execute conceptual matching dealing with context-dependent word ambiguity and produce results in a format that permits the user to interact dynamically to customize and personalized its search strategy.

It is also possible to automate ontology generation and document indexing using the terms similarity based on Conceptual-Latent Semantic Indexing Technique (CLSI). Often time it is hard to find the "right" term and even in some cases the term does not exist.

The ontology is automatically constructed from text document collection and can be used for query refinement. It is also possible to generate conceptual documents similarity map that can be used for intelligent search engine based on CLSI, personalization and user profiling. The user profile is automatically constructed from text document collection and can be used for query refinement and provide suggestions and for ranking the information based on pre-existence user profile.

In our perspective, one can use clarification dialog, user profile, context, and ontology, into an integrated frame work to design a more intelligent search engine. The model will be used for intelligent information and knowledge retrieval through conceptual matching of text. The selected query doesn't need to match the decision criteria exactly, which gives the system a more human-like behavior. The model can also be used for constructing ontology or terms related to the context of search or query to resolve the ambiguity. The new model can execute conceptual matching dealing with context-dependent word ambiguity and produce results in a format that permits the user to interact dynamically to customize and personalized its search strategy.

Given the ambiguity and imprecision of the "concept" in the Internet, which may be described by both textual and image information, the use of Fuzzy Conceptual Model (FCM) (Nikravesh et al., 2003b; Takagi et al., 1995, 1996, 1999a, 1999b) is a necessity for search engines. In the FCM approach, the "concept" is defined by a series of keywords with different weights depending on the importance of each keyword. Ambiguity in concepts can be defined by a set of imprecise concepts. Each imprecise concept in fact can be defined by a set of fuzzy concepts. The fuzzy concepts can then be related to a set of imprecise words given the context. Imprecise words can then be translated into precise words given the ontology and ambiguity resolution through clarification dialog. By constructing the ontology and fine-tuning the strength of links (weights), we could construct a fuzzy set to integrate piecewise the imprecise concepts and precise words to define the ambiguous concept.

3.1 Fuzzy Conceptual Model (FCM)

The central tasks for the most of the search engines can be summarize as 1) query or user information request- do what I mean and not what I say!, 2) model for the Internet, Web representation-web page collection, documents, text, images, music, etc, and 3) ranking or matching function-degree of relevance, recall, precision, similarity, etc. There are two type of search engines that we are interested and are dominating the Internet. First, the most popular search engines that are mainly for unstructured data such as Google ™ and Teoma which are based on the concept of Authorities and Hubs (Figure 1). Second, search engines that are task specifics such as 1) Yahoo!: manually-preclassified, 2) NorthernLight: Classification, 3) Vivisimo: Clustering, 4) Self-organizing Map: Clustering + Visualization and 5) AskJeeves: Natural Languages-Based Search; Human Expert. Google uses the PageRank and Teoma uses HITS for the Ranking. Figure 2 shows the Authorities and Hubs concept and the possibility of comparing two homepages.

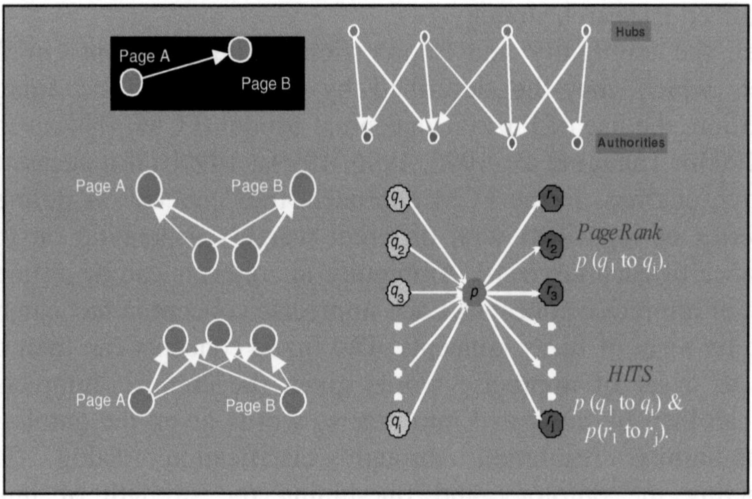

Figure. 2. Similarity of web pages

In this chapter, we will present an intelligent model that can mine the Internet to conceptually match, rank, and cluster the homepages based on predefined linguistic formulations and rules defined by experts or based on a set of known homepages. This model can be used to calculate conceptually the degree of match to the object or query. We will also present the integration of our technology into commercial search engines such as Google ™ as a framework that can be used to integrate our model into any other commercial search engines, or development of the next generation of search engines (Figure 3).

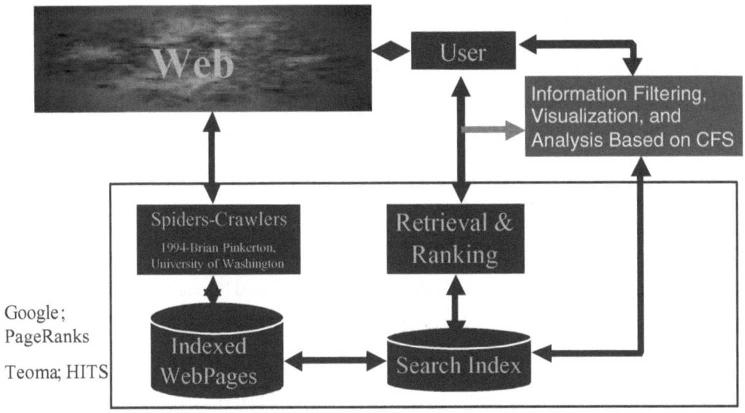

Figure 3. Search Engines Architecture

3.2 Fuzzy Conceptual Model and Search Engine

The Conceptual Fuzzy Match (CFM) model will be used for intelligent information and knowledge retrieval through conceptual matching of both text and images (here defined as "Concept"). The CFM can also be used for constructing fuzzy ontology or terms related to the context of search or query to resolve the ambiguity. It is intended to combine the expert knowledge with soft computing tool. Expert knowledge needs to be partially converted into artificial intelligence that can better handle the huge information stream. In addition, sophisticated management work-flow needs to be designed to make optimal use of this information. In this Chapter, we present the foundation of CFM-Based Intelligent Model and its applications to both information filtering and design of navigation.

In our perspective, one can use clarification dialog, user profile, context, and ontology, into an integrated frame work to design a more intelligent search engine. The model will be used for intelligent information and knowledge retrieval through conceptual matching of text. The selected query doesn't need to match the decision criteria exactly, which gives the system a more human-like behavior. The model can also be used for constructing ontology or terms related to the context of search or query to resolve the ambiguity. The new model can execute conceptual matching dealing with context-dependent word ambiguity and produce results in a format that permits the user to interact dynamically to customize and personalized its search strategy. It is also possible to automate ontology generation and document indexing using the terms similarity based on Conceptual-Latent Semantic Indexing Technique (CLSI). Often time it is hard to find the "right" term and even in some cases the term does not exist. The ontology is automatically constructed from text document collection and can be used for query refinement. It is also possible to generate conceptual documents similarity map that can be used for intelligent search engine based on CLSI, personalization and user profiling. The user profile is automatically constructed from text document collection and can be used for query refinement and provide suggestions and for ranking the information based on pre-existence user profile. Given the ambiguity and imprecision of the "concept" in the Internet, which may be described by both textual and image information, the use of Fuzzy Conceptual Matching (FCM) is a necessity for search engines. In the FCM approach, the "concept" is defined by a series of keywords with different weights depending on the importance of each keyword. Ambiguity in concepts can be defined by a set of imprecise concepts. Each imprecise concept in fact can be defined by a set of fuzzy concepts. The fuzzy concepts can then be related to a set of imprecise words given the context. Imprecise words can then be translated into precise words given the ontology and ambiguity resolution through clarification dialog. By constructing the ontology and fine-tuning the strength of links (weights), we could construct a fuzzy set to integrate piecewise the imprecise concepts and precise words to define the ambiguous concept.

The use of Fuzzy-tf-idf is an alternative to the use of the conventional tf-idf. In this case, the original tf-idf weighting values will be replaced by a fuzzy set rather than original crisp value. To reconstruct such value both ontology and similarity measure can be used. To develop ontology and similarity one can used the conventional Latent Semantic Indexing (LSI) or Fuzzy-LSI [Nikravesh, Azvine (2002)]. The fuzzy-LSI (Figure 4), fuzzy-TF-IDF, and CFS can be used through an integrated system to develop fuzzy conceptual model for intelligent search engine. One can use clarification dialog, user profile, context, and ontology, into a integrated frame work to address some of the issues related to search engines were described earlier. In our perspective, we define this framework as *Fuzzy Conceptual Matching based on Human Mental Model* (Figure 5).

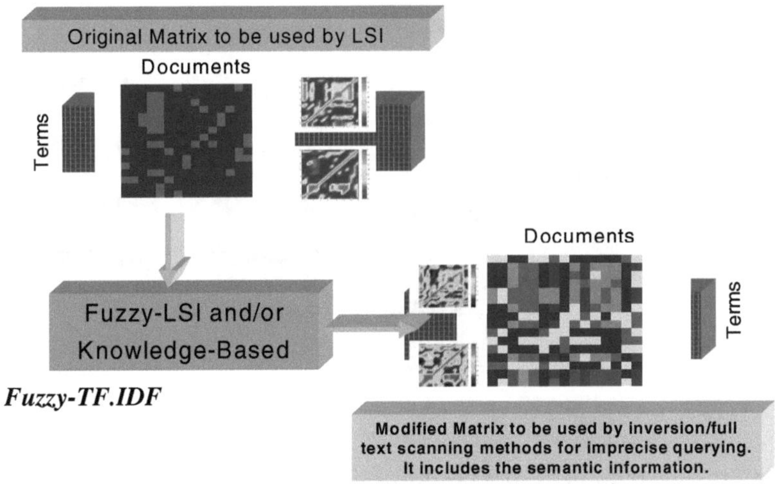

Figure 4. Fuzzy-Latent Semantic Indexing-Based Conceptual Technique.

Figure 6 shows the possible model for similarity analysis is called “fuzzy Conceptual Similarity”. Figure 7 shows the matrix representation of Fuzzy Conceptual Similarity model. Figure 8 shows the evolution of the Term-Document matrix. Figure 9 shows the structure of the Concept-based Google ™ search engine for Multi-Media Retrieval.

There are two type of search engine that we are interested and are dominating the Internet. First, the most popular search engines that are mainly for unstructured data such as Google ™ , MSN, Yahoo! and Teoma which are based on the concept of Authorities and Hubs. Second, search engines that are task spcifics such as 1) Yahoo!: manually-pre-classified, 2) NorthernLight: Classification, 3) Vivisimo: Clustering, 4) Self-organizing Map: Clustering + Visualization and 5) AskJeeves: Natural Languages-Based Search; Human Expert. Google uses the PageRank and Teoma uses HITS (Ding et al. 2001) for the Ranking. To develop such models, NeuSearch™ is needed.

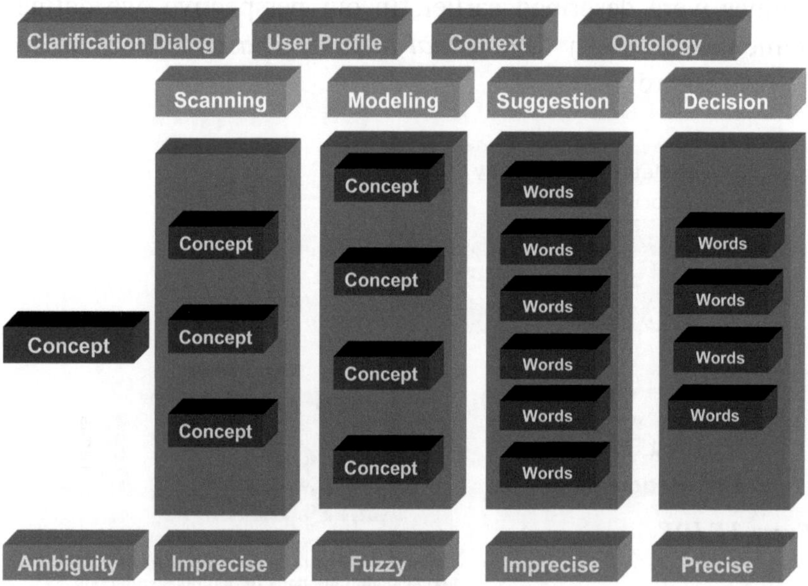

Figure 5. Fuzzy Conceptual Matching and Human Mental Model

3.3 NeuSearch™

To develop such models, NeuSearch™, state-of-the-art computational intelligence techniques are needed (Nikravesh et. al 2002; Nikravesh 2002; Nikravesh and Azvine 2002; Loia et. al 2003). Figures 10 through Figures 13 show, how neuro-science and

PNL can be used to develop the next generation of the search engine. These include and are not limited to:

- Latent-Semantic Indexing and SVD for preprocessing,
- Radial-Basis Function Network to develop concepts,
- Support Vector Machine (SVM) for supervised classification,
- fuzzy/neuro-fuzzy clustering for unsupervised classification based on both conventional learning techniques and Genetic and Reinforcement learning,
- non-linear aggregation operators for data/text fusion,
- automatic recognition using fuzzy measures and a fuzzy integral approach
- self organization map and graph theory for building community and clusters,
- both genetic algorithm and reinforcement learning to learn the preferences,
- fuzzy-integration-based aggregation technique and hybrid fuzzy logic-genetic algorithm for decision analysis, resource allocation, multi-criteria decision-making and multi-attribute optimization.
- text analysis*:* next generation of the Text, Image Retrieval and concept recognition based on soft computing technique and in particular Conceptual Search Model (CSM). This includes
    - Understanding textual content by retrieval of relevant texts or paragraphs using CSM followed by clustering analysis.
    - Hierarchical model for CSM
    - Integration of Text and Images based on CSM
    - CSM Scalability, and
    - The use of CSM for development of
        - Ontology
        - Query Refinement and Ambiguity Resolution
        - Clarification Dialog
        - Personalization-User Profiling

Figure 10 shows a unified framework for development of search engine based on conceptual semantic indexing. This model will be used to develop the NeuSearch model based on Neuroscience and PNL approach. As explained in earlier section and represented by **Figures 6** through **8** with respect to development of FCM, the first step will be to represent the term-document matrix. Tf-idf is the starting point for our model.

Once fuzzy tf-idf (term-document matrix) is created, the next step will be to use such indexing to develop the search and information retrieval mechanism. As shown in **Figure 10,** there are two alternatives 1) classical search models such as LSI-based approach and 2) NeuSearch model. The LSI based models include 1) Probability based-LSI, Bayesian-LSI, Fuzzy-LSI, and NNNet-based-LSI models. It is interesting enough that one can find an alternative to each of such LSI-based models using Radial Basis Function (RBF). For example, one can use Probabilistic RBF equivalent to Probability based-LSI, Generalized RBF as an equivalent to Bayesian-LSI, ANFIS as an equivalent to Fuzzy-LSI, and RBF function neural network (RBFNN) as an equivalent to NNnet-LSI. RBF model is the basis of the NeuSearch model (Neuro-Fuzzy Conceptual Search-- NeuFCS). Given the NeuSearch model, one needs to calculate the w(i,j) which will be defined in next section (the network weights). Depends on the model and framework used (Probability, Bayesian, Fuzzy, or NNnet model), the interpretation of the weights will be different. **Figure 11** shows typical Neuro-Fuzzy Conceptual Search (NeuFCS) model input-output. The main and core idea with respect to NeuSearch is the fact that human does mainly search given concept and context of his/her interest rather than in absolute form. For example, one can search for word "Java" in the context of "Computer" or in the context of "Coffee". One can also search for "Apple" in the context of "Fruit" or in the context of "Computer". Therefore, before we do relate the terms to document, we first extend the keyword, given the existing concepts-contexts and then we do relate that term to the documents. Therefore, there will be always a two steps process, based on NeuSearch model as shown below:

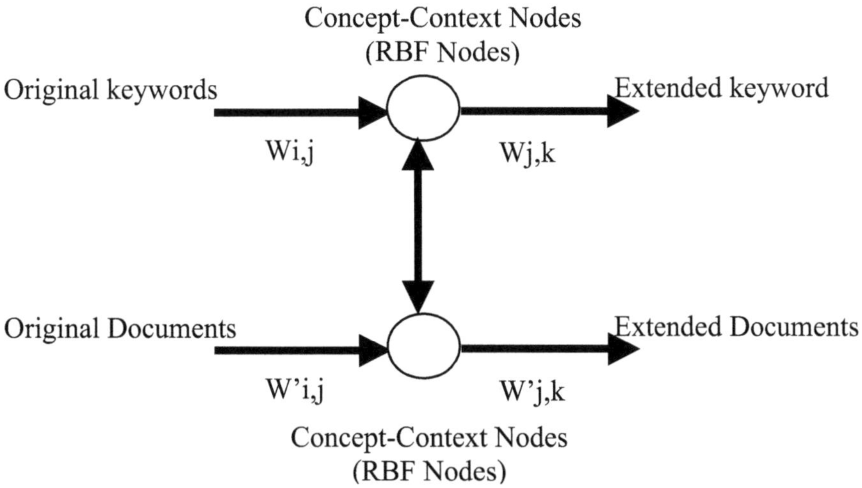

In general, w(i,j) is function of p(i,j), p(i), and p(j), where p represents the probability. Therefore, one can use the probabilistic-LSI or PRBF given the NeuSearch framework. If the probabilities are not known, which often times is the case, one can use Fuzzy-LSI model or ANFIS model or NNnet-LSI or RBFNN model using the NeuSearch model. In general, the PNL model can be used as unified framework as shown in **Figure 12**. **Figure 12** shows PNL-Based Conceptual Fuzzy Search Using Brain Science model and concept presented based on **Figures 6** through **8 and Figures 10** through **12.**

Based on PNL approach, w(i,j) is defined based on ri,j as follows:

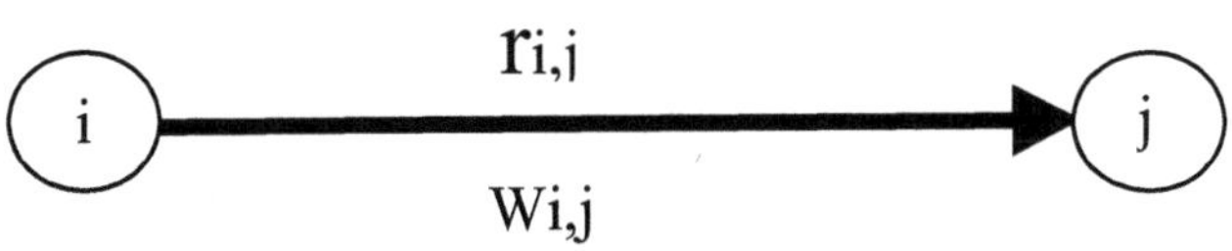

Where wi,j is granular strength of association between i and j, ri,j is epistemic lexicon, wi,j <== ri,j, and ri,j is defined as follows:

rij: ***i is an instance of j*** ***(is or isu)***

i is a subset of j ***(is or isu)***
i is a superset of j ***(is or isu)***
j is an attribute of i
i causes j ***(or usually)***
i and j are related

Figure 13 shows typical example of Neu-FCS result. Both term-document matrix and term-term document matrix/document-document matrix are reconstructed. For example, the original document corpus is very sparse as one can see in the figure; however, after processing through concept node, the document corpus is less spare and smoothens. Also, the original keywords are expanded given the concept-context nodes. While the nodes in the hidden space (Concept-Context nodes) are based on RBF (NeuSearch model) and it is one dimensional nodes, one can use SOM to rearrange the nodes given the similarity of the concepts (in this case the RBF in each node) and present the 1-D nodes into 2-D model as shown in **Figure 13.**

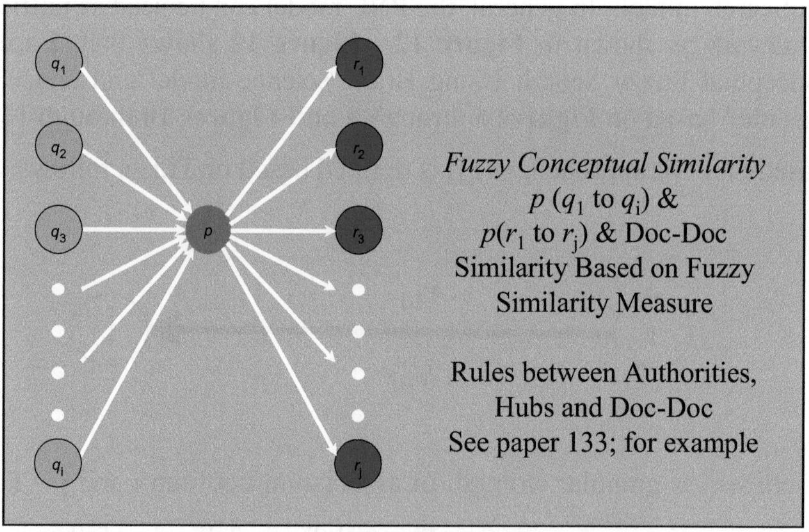

Figure 6. Fuzzy Conceptual Similarity

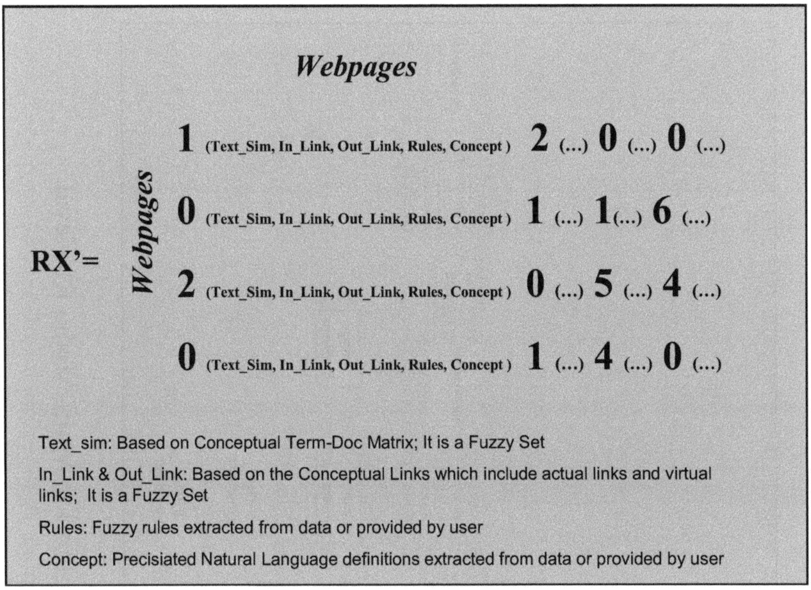

Figure 7. Matrix representation of Fuzzy Conceptual Similarity model

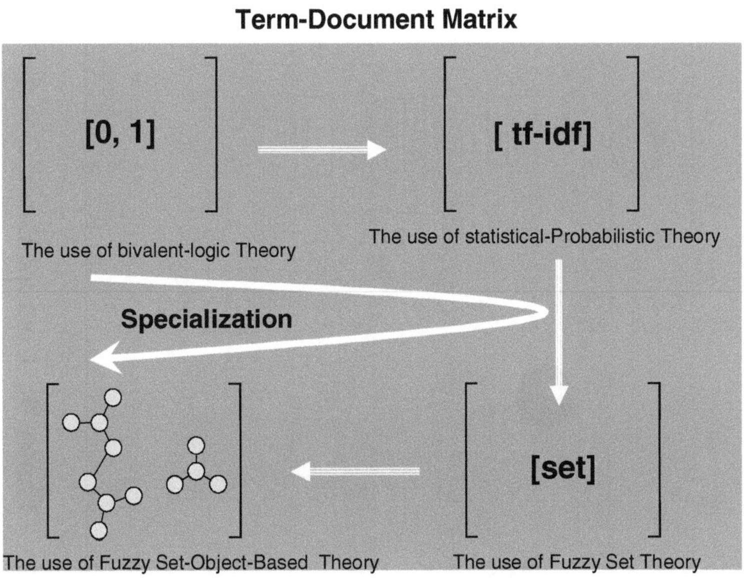

Figure 8. Evolution of Term-Document Matrix representation

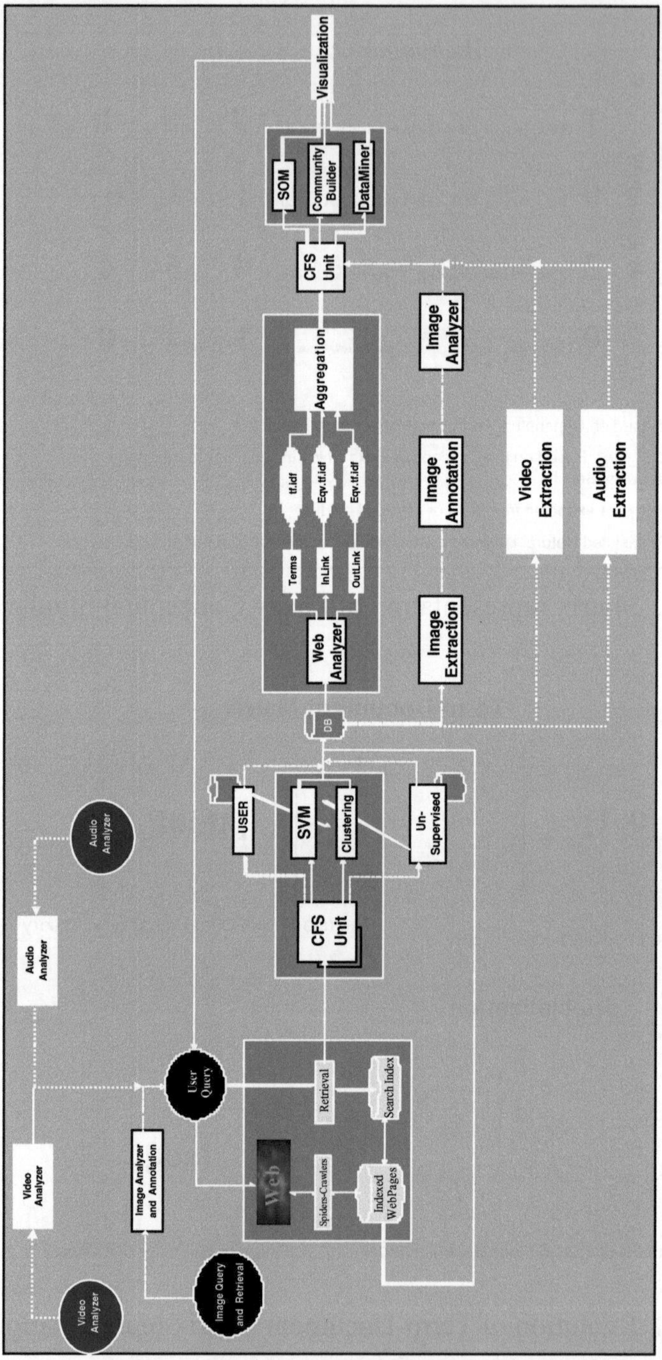

Figure 9. Concept-Based Google™ Search Engine for Multi-Media Retrieval

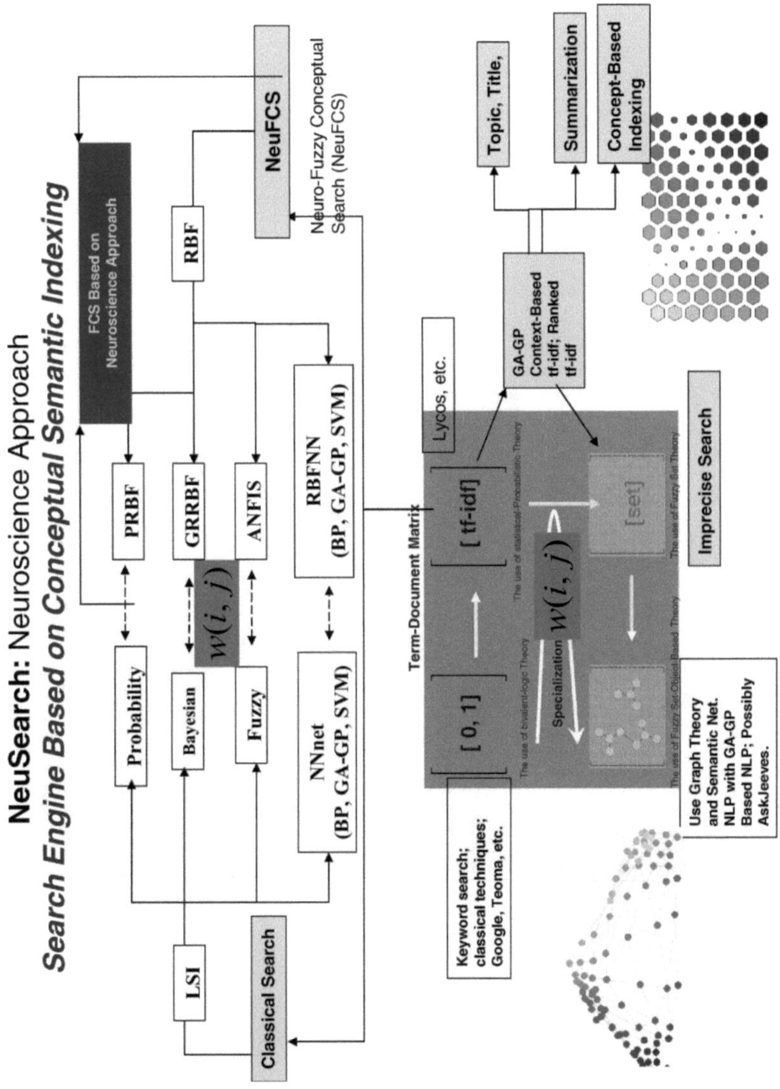

Figure 10. Search engine based on conceptual semantic indexing

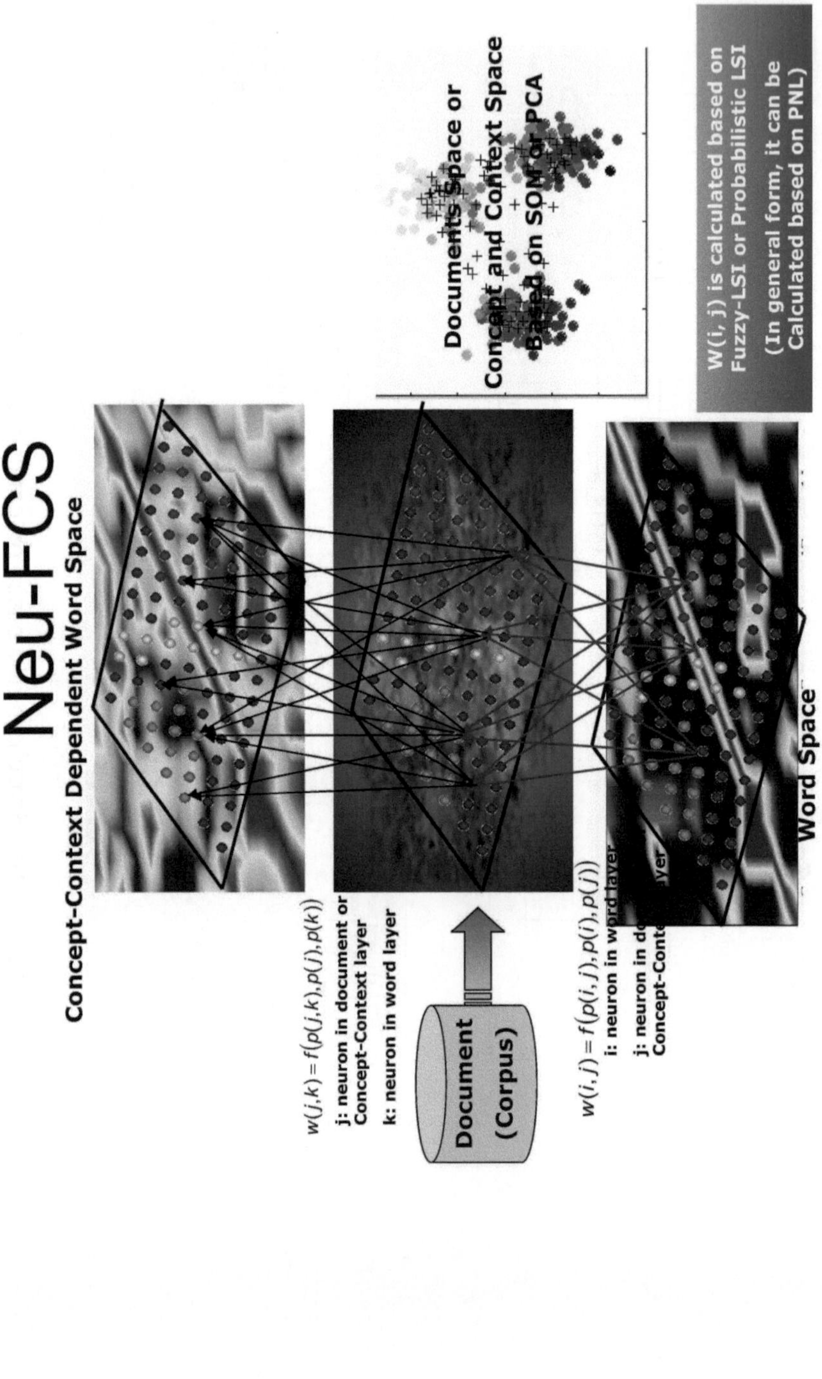

Figure 11. Neuro-Fuzzy Conceptual Search (NeuFCS)

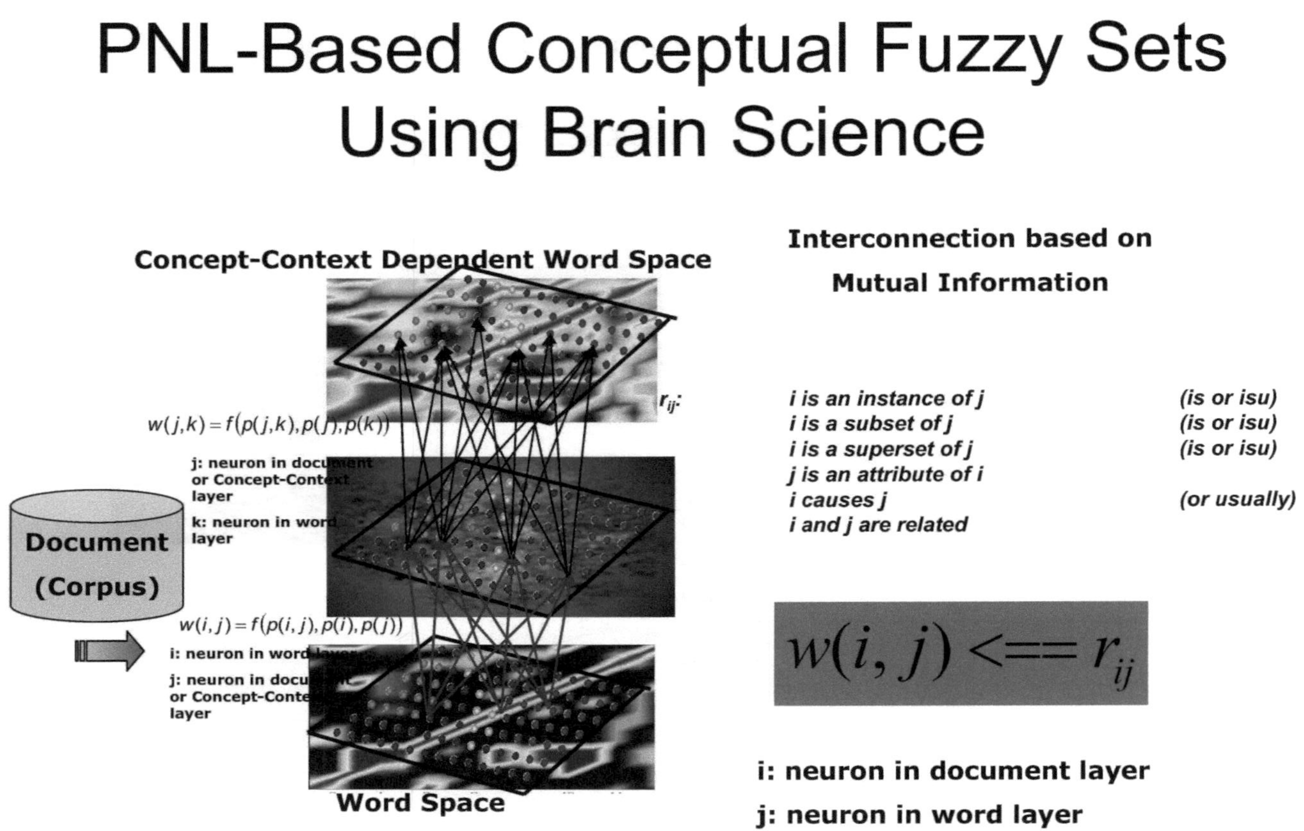

Figure 12. PNL-Based Conceptual Fuzzy Sets Using Brain Science

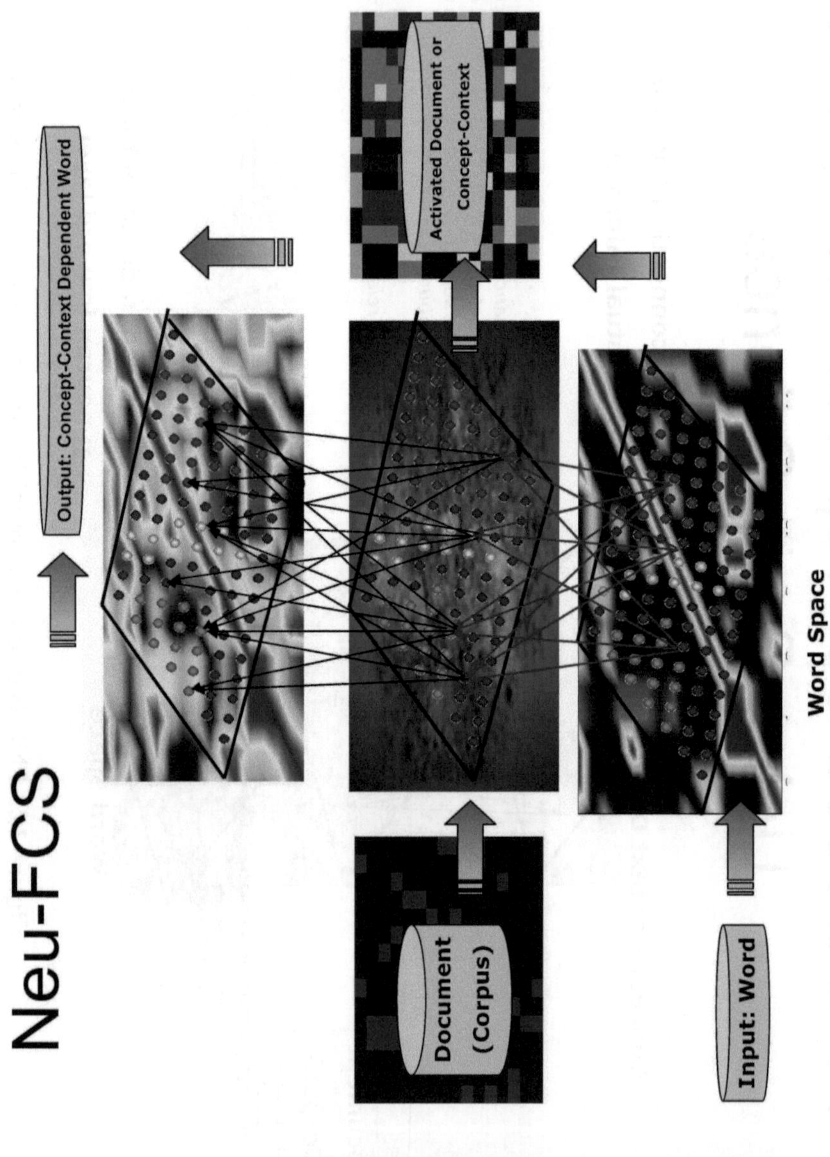

Figure 13. Typical example of Neu-FCS

4. From Search Engine to Q/A Systems: The Need for New Tools (Nikravesh et al., Web Intelligence: Conceptual-Based Model, Memorandum No. UCB/ERL M03/19, 5 June 2003)

Construction of Q/A systems has a long history in AI. Interest in Q/A systems peaked in the seventies and eighties, and began to decline when it became obvious that the available tools were not adequate for construction of systems having significant question-answering capabilities. However, Q/A systems in the form of domain-restricted expert systems have proved to be of value, and are growing in versatility, visibility and importance.

Search engines as we know them today owe their existence and capabilities to the advent of the Web. A typical search engine is not designed to come up with answers to queries exemplified by "How many Ph.D. degrees in computer science were granted by Princeton University in 1996?" or "What is the name and affiliation of the leading eye surgeon in Boston?" or "What is the age of the oldest son of the President of Finland?" or "What is the fastest way of getting from Paris to London?"

Upgrading a search engine to a Q/A system is a complex, effort-intensive, open-ended problem. Semantic Web and related systems for upgrading quality of search may be viewed as steps in this direction. But what may be argued, as is done in the following, is that existing tools, based as they are on bivalent logic and probability theory, have intrinsic limitations. The principal obstacle is the nature of world knowledge.

The centrality of world knowledge in human cognition, and especially in reasoning and decision-making, has long been recognized in AI. The Cyc system of Douglas Lenat is a repository of world knowledge. The problem is that much of world knowledge consists of perceptions. Reflecting the bounded ability of sensory organs, and ultimately the brain, to resolve detail and store information, perceptions are intrinsically imprecise. More specifically, perceptions are f-granular in the sense that (a) the boundaries of perceived classes are fuzzy; and (b) the perceived values of attributes are granular, with a granule being a clump of values drawn together by indistinguishability, similarity, proximity or functionality. What is not widely recognized is that f-granularity of perceptions put them well beyond the reach of computational bivalent-logic-based theories. For example, the meaning of a simple perception described as "Most Swedes are tall," does not admit representation in predicate logic and/or probability theory.

Dealing with world knowledge needs new tools. A new tool which is suggested for this purpose is the fuzzy-logic-based method of computing with words and perceptions (CWP) (Zadeh 1999, 2001, 2002; Zadeh and Nikravesh 2002)with the understanding that perceptions are described in a natural language (Nikravesh and Azvine 2001; Loia et. al 2002; Nikravesh et. al 2003; Nikravesh 2001a, 2001b). A concept which plays a key role in CWP is that of Precisiated Natural Language (PNL). It is this language that is the centerpiece of our approach to reasoning and decision-making with world knowledge.

A concept which plays an essential role in PNL is that of precisiability. More specifically, a proposition, p, in a natural language, NL, is PL precisiable, or simply precisiable, if it is translatable into a mathematically well-defined language termed precisiation language, PL. Examples of precisiation languages are: the languages of propositional logic; predicate logic; modal logic; etc.; and Prolog; LISP; SQL; etc. These languages are based on bivalent logic. In the case of PNL, the precisiation language is a fuzzy-logic-based language referred to as the Generalized Constraint Language (GCL). By construction, GCL is maximally expressive.

A basic assumption underlying GCL is that, in general, the meaning of a proposition, p, in NL may be represented as a generalized constraint of the form X isr R, where X is the constrained variable; R is the constraining relation, and r is a discrete-valued variable, termed modal variable, whose values define the modality of the constraint, that is, the way in which R constrains X. The principal modalities are; possibilistic (r=blank); probabilistic (r=p); veristic (r=v); usuality (r=u); fuzzy random set (r=rs); fuzzy graph (r=fg); and Pawlak set (r=ps). In general, X, R and r are implicit in p. Thus, precisiation of p, that is, translation of p into GCL, involves explicitation of X, R and r. GCL is generated by (a) combining generalized constraints; and (b) generalized constraint propagation, which is governed by the rules of inference in fuzzy logic. The translation of p expressed as a generalized constraint is referred to as the GC-form of p, GC(p). GC(p) may be viewed as a generalization of the concept of logical form. An abstraction of the GC-form is referred to as a protoform (prototypical form) of p, and is denoted as PF(p). For example, the protoform of p: "Most Swedes are tall" is Q A's are B's, where A and B are labels of fuzzy sets, and Q is a fuzzy quantifier. Two propositions p and q are said to be PF-equivalent if they have identical protoforms. For example, "Most Swedes are tall," and "Not many professors are rich," are PF-equivalent. In effect, a protoform of p is its deep semantic structure. The protoform language, PFL, consists of protoforms of elements of GCL.

With the concepts of GC-form and protoform in place, PNL may be defined as a subset of NL which is equipped with two dictionaries: (a) from NL to GCL; and (b) from GCL to PFL. In addition, PNL is equipped with a multiagent modular deduction database, DDB, which contains rules of deduction in PFL. A simple example of a rule of deduction in PFL which is identical to the compositional rule of inference in fuzzy logic, is: if X is A and (X, Y) is B then Y is A∘B, where A∘B is the composition of A and B, defined by $\mu_B(v) = \sup_u(\mu_A(u) \wedge \mu_B(u,v))$, *where* μ_A *and* μ_B *are the membership functions of A and B, respectively, and* $\wedge$ *is min or, more generally, a T-norm. The rules of deduction in DDB are organized into modules and submodules, with each module and submodule associated with an agent who controls execution of rules of deduction and passing results of execution.*

In our approach, PNL is employed in the main to represent information in the world knowledge database (WKD). For example, the items:

If X/Person works in Y/City then it is likely that X lives in or near Y
If X/Person lives in Y/City then it is likely that X works in or near Y
are translated into GCL as:

Distance (Location (Residence (X/Person), Location (Work (X/Person) isu near, where isu, read as ezoo, is th e usuality constraint. The corresponding protoform is:

F (A(B(X/C), A(E(X/C)) isu G.

A concept which plays a key role in organization of world knowledge is that of an epistemic (knowledge-directed) lexicon (EL). Basically, an epistemic lexicon is a network of nodes and weighted links, with node i representing an object in the world knowledge database, and a weighted link from node i to node j representing the strength of association between i and j. The name of an object is a word or a composite word, e.g., car, passenger car or Ph.D. degree. An object is described by a relation or relations whose fields are attributes of the object. The values of an attribute may be granulated and associated with granulated probability and possibility distributions. For example, the values of a granular attribute may be labeled small, medium and large, and their probabilities may be described as low, high and low, respectively. Relations which are associated with an object serve as PNL-based descriptions of the world knowledge about the object. For example, a relation associated with an object labeled Ph.D. degree may contain attributes labeled Eligibility, Length.of.study, Granting.institution, etc. The knowledge associated with an object may be context-dependent. What should be stressed is that the concept of an epistemic lexicon is intended to be employed in representation of world knowledge — which is largely perception- based—rather than Web knowledge, which is not.

As a very simple illustration of the use of an epistemic lexicon, consider the query "How many horses received the Ph.D. degree from Princeton University in 1996." No existing search engine would come up with the correct answer, "Zero, since a horse cannot be a recipient of a Ph.D. degree." To generate the correct answer, the attribute Eligibility in the Ph.D. entry in EL should contain the condition "Human, usually over twenty years of age."

In conclusion, the main thrust of the fuzzy-logic-based approach to question-answering which is outlined here, is that to achieve significant question-answering capability it is necessary to develop methods of dealing with the reality that much of world knowledge—

and especially knowledge about underlying probabilities is perception-based. Dealing with perception-based information is more complex and more effort-intensive than dealing with measurement-based information. In this instance, as in many others, complexity is the price that has to be paid to achieve superior performance.

5. BISC-Decision Support System and Intelligent Information Systems in Enterprise

In this section, a web-based model to decision model for analysis of structured database will be presented.

Most of the existing search systems' 'software' are modeled using crisp logic and queries. We introduce fuzzy querying and ranking as a flexible tool allowing approximation where the selected objects do not need to match exactly the decision criteria resembling natural human behavior. The model consists of five major modules: the Fuzzy Search Engine, the Application Templates, the User Interface, the Database and the Evolutionary Computing. The system is designed in a generic form to accommodate more diverse applications and to be delivered as stand-alone software to academia and businesses.

Searching database records and ranking the results based on multi-criteria queries is central for many database applications used within organizations in finance, business, industrial and other fields. Most of the available systems' 'software' are modeled using crisp logic and queries, which result in rigid systems with imprecise and subjective processes and results. In this chapter we introduce fuzzy querying and ranking as a flexible tool allowing approximation where the selected objects do not need to match exactly the decision criteria resembling natural human behavior. The model consists of five major modules: the Fuzzy Search Engine (FSE), the Application Templates (AT), the User Interface (UI), the Database (DB) and the Evolutionary Computing (EC). We developed the software with many essential features. It is built as a web-based software system that users can access and use over the Internet. The system is designed to be generic so that it can run different application domains. To this end, the Application Template module provides information of a specific application as attributes and properties, and serves as a guideline structure for building a new application. The Fuzzy Search Engine (FSE) is the core module of the system. It has been developed to be generic so that it would fit any application. The main FSE component is the query structure, which utilizes membership functions, similarity functions and aggregators. Through the user interface a user can enter and save his profile, input criteria for a new query, run different queries and display results. The user can manually eliminate the results he disapproves of or change the ranking according to his preferences. The Evolutionary Computing (EC) module monitors ranking preferences of the user's queries. It learns to adjust to the intended meaning of the users' preferences.

We present our approach with three important applications: ranking (scoring) (Nikravesh, 2001a, 2001b, 2001c), which has been used to make financing decisions concerning credit cards, cars and mortgage loans; the process of college admissions where hundreds of thousands of applications are processed yearly by U.S. universities; and date matching as one of the most popular internet programs. Even though we implemented three applications, the system is designed in a generic form to accommodate more diverse applications and to be delivered as stand-alone software to academia and businesses.

5.1 BISC Decision Support System

The BISC (Berkeley Initiative in Soft Computing) Decision Support System Components include:

- Data Management
  - database(s) which contains relevant data for the decision process
- User Interface
  - users and DSS communication
- Model Management and Data Mining

- o includes software with quantitative and fuzzy_models including aggregation process, query, ranking, and fitness evaluation
- Knowledge Management and Expert System
  - o model representation including linguistic formulation
- Evolutionary Kernel and Learning Process
- Data Visualization and Visual Interactive Decision Making

5.2 Model Framework

The DSS system starts by loading the application template, which consists of various configuration files for a specific application and initializing the database for the application, before handling a user's requests, see Figure 14. Once the DSS system is initialized, users can enter their own profiles in the user interface or make a search

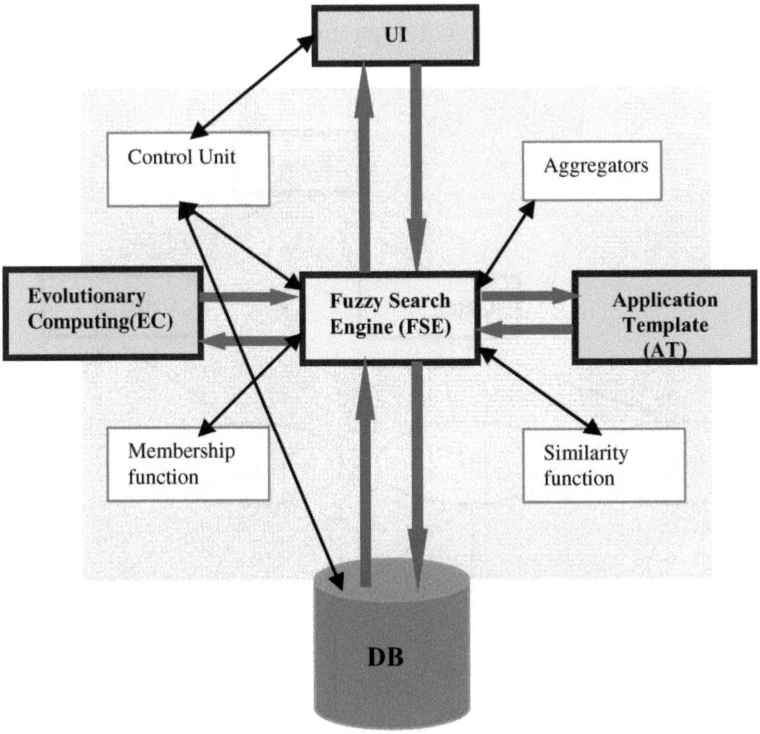

Figure 14. The BISC-DSS general framework

with their preferences. These requests are handled by the control unit of the system. The control unit converts user input into data objects that are recognized by the DSS system. Based on the request types, it forwards them to the appropriate modules. If the user wants to create a profile, the control unit will send the profile data directly to the database module which stores the data in the database for the application.

If the user wants to query the system, the control unit will direct the user's preferences to the Fuzzy Search Engine which queries the database. The query results will be sent back to the control unit and displayed to the users.

5.2.1 Fuzzy Engine

Fuzzy Query, search and Ranking: To support generic queries, the fuzzy engine has been designed to have a tree structure. There are two types of nodes in the tree, category nodes and attribute nodes, as depicted in Figure 15.

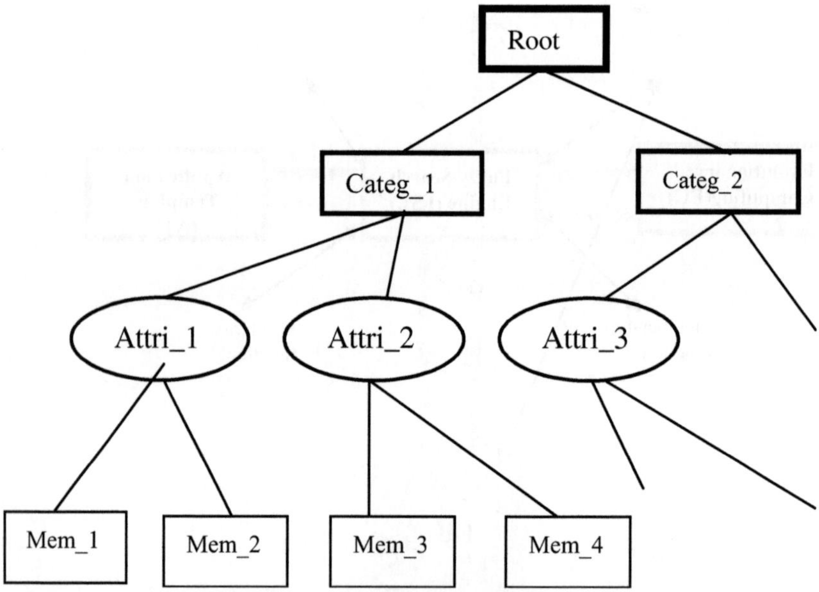

Figure 15. The Fuzzy search engine tree structure.

While multiple category levels are not necessary, they are designed to allow various refinements of the query through the use of the type of aggregation of the children. Categories only act to aggregate the lower levels. The attribute nodes contain all the important information about a query. They contain the membership functions for the fuzzy comparison as well as the use of the various aggregation methods to compare two values. The attribute nodes handle the compare method slightly differently than the category nodes. There are two different ways attributes may be compared. The attribute nodes contain a list of membership functions comprising the fuzzy set. The degrees of membership for this set are passed to the similarity comparator object, which currently has a variety of different methods to calculate the similarity between the two membership vectors. In the other method, the membership vector created by having full membership to a single membership function specified in the fuzzy data object, but no membership value for the other functions.

Membership function: Currently there are three membership functions implemented for the Fuzzy Engine. A generic interface has been created to allow several different types of membership functions to be added to the system. The three types of membership functions in the system are: Gaussian, Triangular and Trapezoidal. These functions have three main points, for the lower bound, upper bound and the point of maximum membership. For other functions, optional extra points may be used to define the shape (an extra point is required for the trapezoidal form).

5.2.2 Application Template

The DSS system is designed to work with different application domains. The application template is a format for any new application we build; it contains data of different categories, attributes and membership functions of that application. The application template module consists of two parts: 1) the application template data file specifies all the membership functions, attributes and categories of an application. We can consider it as a configuration data file for an application. It contains the definition of membership functions, attributes and the relationship between them;

2) The application template logic parses and caches data from the data file so that other modules in the system can have faster access to definitions of membership functions, attributes and categories. It also creates a tree data structure for the fuzzy search engine to transverse.

5.2.3 User Interface

It is difficult to design a generic user interface that suits different kind of applications for all the fields. For example, we may want to have different layouts for user interfaces for different applications. To make the DSS system generic while preserving the user friendliness of the interfaces for different applications, we developed the user interfaces into two parts (see Figure 16).

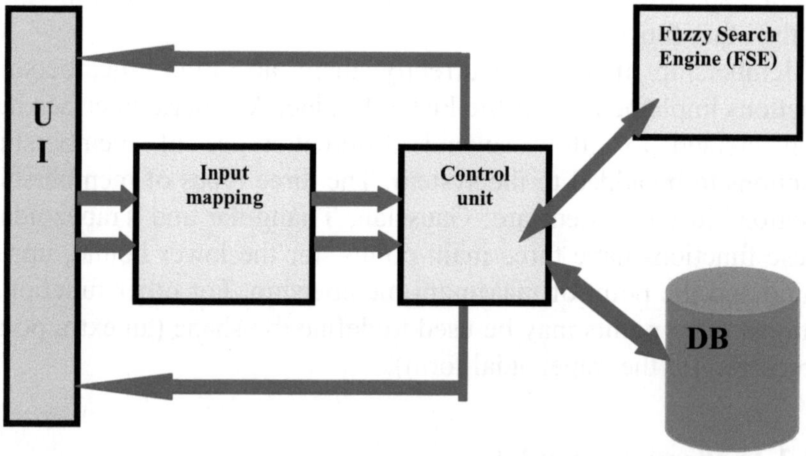

Figure 16. User interface data flow

First, we designed a specific HTML interface for each application we developed. Users can input their own profiles; make queries by specifying preferences for different attributes. Details for the DSS system are encapsulated from the HTML interface so that the HTML interface design would not be constrained by the DSS system. The second part of our user interface module is a mapping between the parameters in the HTML files and the attributes in the application template module for the application. The input mapping specifies the attribute names to which each parameter in the HTML interface corresponds.. With this input mapping, a user interface designer can use input methods and parameter names freely.

5.2.4 Database (DB)

The database module is responsible for all the transactions between the DSS system and the database. This module handles all queries or user profile creations from the Fuzzy Engine and the Control Unit respectively. For queries from the Fuzzy Search Engine, it retrieves data from the database and returns it in a data object form. Usually queries are sets of attribute values and their associated weights. The database module returns the matching records in a format that can be manipulated by the user, such as eliminating one or more record or changing their order. To create a user profile, it takes data objects from the Control Unit and stores it in the database. There are three components in the DB module (see Figure 17).

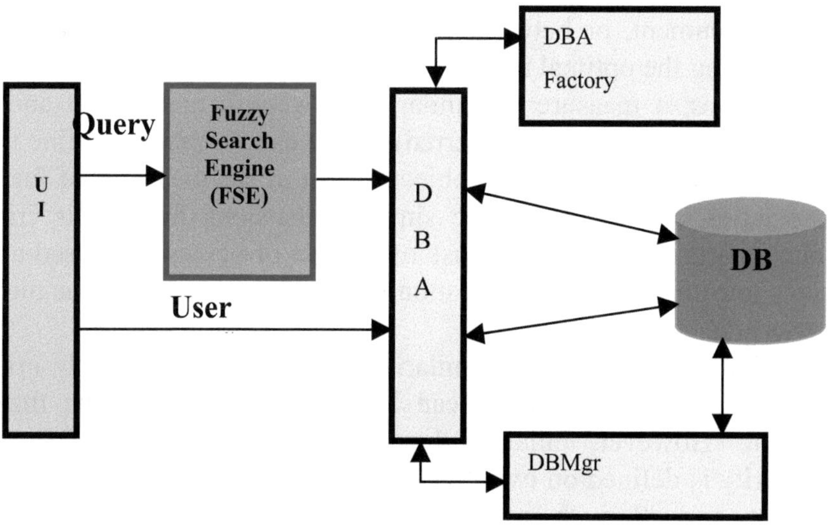

Figure 17. Database module components

The DB Manager is accountable for two things: 1) setting up database connections and allocating database connections to DB Accessor objects when needed. It also supplies information to the database for authentication purposes (e.g. username, password, path to the database etc); 2) The DB Accessor Factory creates DB Accessor objects for a specific application. For example, if the system is running the date matching application, DB Accessor Factory will create DB Accessor objects for the date matching application. The existence of this class serves the purpose of using a generic Fuzzy Search Engine; 3) the DB Accessor is responsible for storing and getting user profiles to and from the database. It is the component that queries the database and wrap result from the database into data objects that are recognized by our application framework.

5.2.5 Measure of Association and Fuzzy Similarity

As in crisp query and ranking, an important concept in fuzzy query and ranking applications is the measure of association or similarity between two objects in consideration. For example, in a fuzzy query application, a measure of similarity between two queries and a document, or between two documents, provides a basis for determining the optimal response from the system. In fuzzy ranking applications, a measure of similarity between a new object and a known preferred (or non-preferred) object can be used to define the relative goodness of the new object. Most of the measures of fuzzy association and similarity are simply extensions from their crisp counterparts. However, because of the use of perception based and fuzzy information, the computation in the fuzzy domain can be more powerful and more complex.

Various definitions of similarity exist in the classical, crisp domain, and many of them can be easily extended to the fuzzy domain. However, unlike in the crisp case, in the fuzzy case the similarity is defined on two fuzzy sets. Suppose we have two fuzzy sets A and B with membership functions $\mu_A(x)$ and $\mu_B(x)$, respectively. The arithmetic operators involved in the fuzzy similarity measures can be treated using their usual definitions while the union and the intersection operators need to be treated specially. It is important for these operator pairs to have the following properties: (1) conservation, (2) monotonicity, (3) commutativity, and (4) associativity. It can be verified that the triangular norm (T-norm) and triangular co-norm (T-conorm) (Nikravesh, 2001a; Bonissone and Decker, 1986; Mizumoto, 1989; Fagin, 1998 and 1999) conform to these properties and can be applied here. A detailed survey of some commonly used T-norm and T-conorm pairs along with other aggregation operators can be find at Nikravesh and Azvine (2002) .

In many situations, the controlling parameters, including the similarity metric, the type of T-norm/conorm, the type of aggregation operator and associated weights, can all be specified based on the domain knowledge of a particular application. However, in some other cases, it may be difficult to specify a priori an optimal set of parameters. In those cases, various machine learning methods can be employed to automatically "discover" a suitable set of parameters using a supervised or unsupervised approach. For example, the Genetic Algorithm (GA) and DNA-based computing, as described in later sections, can be quite effective.

5.3 Evolutionary Computing

In the Evolutionary Computing (EC) module of the BISC Decision Support System, our purpose is to use an evolutionary method to allow automatic adjusting of the user's preferences. These preferences can be seen as parameters of the fuzzy logic model in form of weighting of the used variables. These preferences are then represented by a weight vector and genetic algorithms will be used to fix them.

In the Evolutionary Computation approach, Genetic Programming, which is an extension of Genetic Algorithms, is the closest technique to our purpose. It allows us to learn a tree structure, which represents the combination of aggregators. The selection of these aggregators is included to the learning process using the Genetic Programming.

In this section, we describe the GA (Holland, 1992) and GP (Koza, 1992) application to our problem. Our aim is learning fuzzy-DSS parameters which are the weight vectors representing the user preferences associated to the variables that have to be aggregated on the one hand, and the adequate decision tree representing the combination of the aggregation operators that have to be used on the other hand.

Weight vector being a linear structure, can be represented by a binary string in which weight values are converted to binary numbers. This binary string corresponds to the individual's DNA in the GA learning process. The goal is to find the optimal weighting of the variables. A general GA module can be used by defining a specific fitness function for each application as shown in **Figure 18.**

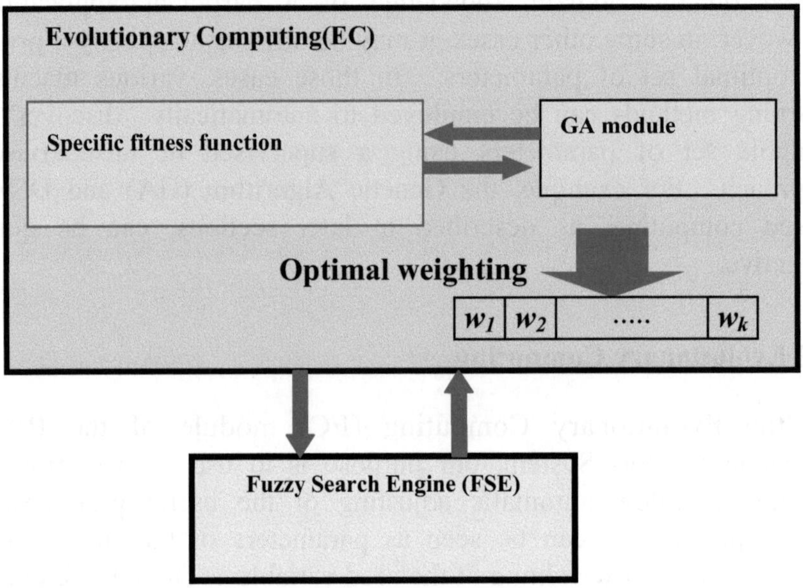

Figure 18. Evolutionary Computing Module: preferences learning

Aggregators can be combined in the form of a tree structure which can be built using a Genetic Programming learning module (Figure 19). It consists in evolving a population of individuals represented by tree structures. The evolution principle remains the same as in a conventional GP module but the DNA encoding needs to be defined according to the considered problem. We propose to define an encoding for aggregation trees which is more complex than for classical trees and which is common to all considered applications. As shown in Figure 19, we need to define a specific encoding, in addition to the fitness function specification.

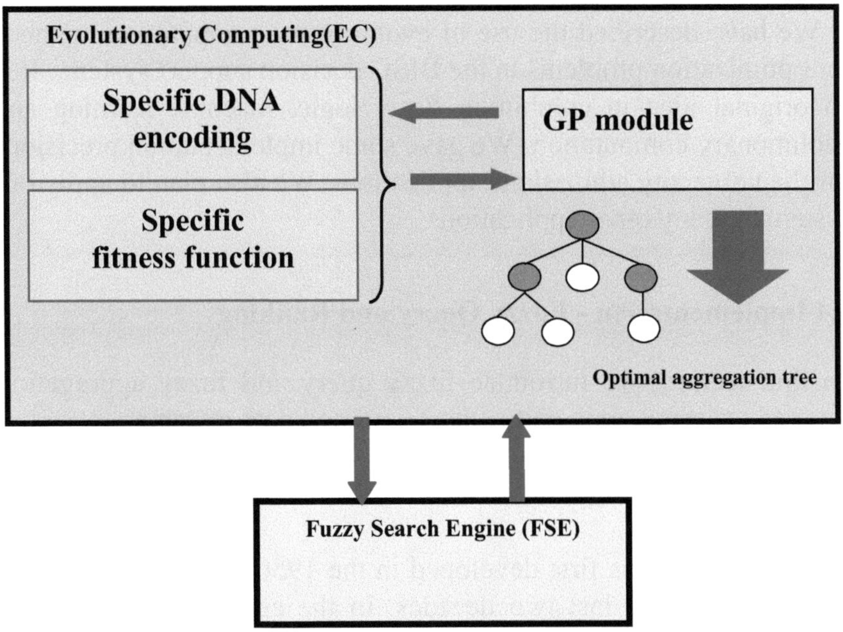

Figure 19. Evolutionary Computing Module: aggregation tree learning

Tree structures are generated randomly as in the conventional GP. But, since these trees are augmented according the properties defined above, the generation process has to be updated. So, we decided to randomly generate the number of arguments when choosing an aggregator as a node in the tree structure. And for the weights, we chose to generate them randomly for each node during its creation.

Concerning the fitness function, it is based on performing the aggregation operation and the root node of the tree that has to be evaluated. For the university admissions application, the result of the root execution corresponds to the score that has to be computed for each value vector in the training data set. The fitness function, as in the GA learning of the user preferences, consists in simple or combined similarity measures. In addition, we can include to the fitness function a complementary measure that represents the individual's size which has to be minimized in order to avoid over-sized trees.

We have described the use of evolutionary computation methods for optimization problems in the BISC decision support system. It is an original idea in combining fuzzy logic, machine learning and evolutionary computation. We gave some implementation precisions for the university admissions application. We also plan to apply our system to many other applications.

5.4 Implementation - Fuzzy Query and Ranking

In this section, we introduce fuzzy query and fuzzy aggregation for credit scoring, university admissions and date matching.

5.4.1 Credit Scoring

Credit scoring was first developed in the 1950's and has been used extensively in the last two decades. In the early 1980's, the three major credit bureaus, Equitax, Experian, and TransUnion worked with the Fair Isaac Company to develop generic scoring models that allow each bureau to offer an individual score based on the contents of the credit bureau's data. When you apply for financing, whether it's a new credit card, car or student loan, or a mortgage, about 40 pieces of information from your credit card report are fed into a model (Nikravesh and Azvine 2002 and Nikravesh et al. 2003). This information is categorized into the following five categories with different levels of importance (% of the score):

Past payment history (35%)
Amount of credit owed (30%)
Length of time credit established (15%)
Search for and acquisition of new credit (10%)
Types of credit established (10%)

Given the factors presented earlier, a simulated model has been developed. A series of excellent, very good, good, not good, not bad, bad, and very bad credit scores have been recognized (without including history). Then, fuzzy similarity and ranking have been used to rank the new user and define his/her credit score.. In the inference engine, the rules based on factual knowledge (data) and knowledge drawn from human experts (inference) are combined, ranked, and clustered based on the confidence level of human and factual support. This information is then used to build the fuzzy query model with associated weights. In the query level, an intelligent knowledge-based search engine provides a means for specific queries. Initially we blend traditional computation with fuzzy reasoning. This effectively provides validation of an interpretation, model, hypothesis, or alternatively, indicates the need to reject or reevaluate. Information must be clustered, ranked, and translated to a format amenable to user interpretation.

5.4.2 University Admissions

Hundreds of millions of applications were processed by U.S. universities resulting in more than 15 million enrollments in the year 2000 for a total revenue of over $250 billion. College admissions are expected to reach over 17 million by the year 2010, for total revenue of over $280 billion.

The UC Berkeley campus admits its freshman class on the basis of an assessment of the applicants' high school academic performance (approximately 50%) and through a comprehensive review of the application including personal achievements of the applicant (approximately 50%) (University of California-Berkeley).

At Stanford University, in addition to the academic transcript, close attention is paid to other factors such as student's written application, teacher references, the short responses and one-page essay (carefully read for quality, content, and creativity), and personal qualities.

Given the factors and general admission ***criteria***, a simulated-hypothetical model (a Virtual Model) was developed. A series of excellent, very good, good, not good, not bad, bad, and very bad student given the criteria for admission has been recognized. These criteria over time can be modified based on the success rate of students admitted to the university and their performances during the first, second, third and fourth years of their education with different weights and degrees of importance given for each year. Then, fuzzy similarity and ranking can evaluate a new student rating and find it's similarity to a given set of criteria.This will provide the ability to answer "what if" questions in order to decrease uncertainty and provide a better risk analysis to improve the chance for "increased success" on student selection or it can be used to select students on the basis of "diversity" criteria. The model can be used as for decision support and for a more uniform, consistent and less subjective and biased way. Finally, the model could learn and provide the mean to include the feedback into the system through time and will be adapted to the new situation for defining better criteria for student selection.

In this study, it has been found that ranking and scoring is a very subjective problem and depends on user perception and preferences in addition to the techniques used for the aggregation process which will effect the process of the data mining in reduced domain. Therefore, user feedback and an interactive model are recommended tools to fine-tune the preferences based on user constraints. This will allow the representation of a multi-objective optimization with a large number of constraints for complex problems such as credit scoring or admissions. To solve such subjective and multi-criteria optimization problems, GA-fuzzy logic and DNA-fuzzy logic models are good candidates.To solve such subjective and multi-criteria optimization problems with a large number of constraints for complex problems such as university admissions, the BISC Decision Support System is an excellent candidate.

5.4.3 Date Matching

The main objective is to find the best possible match in the huge space of possible outputs in the databases using the imprecise matching such as fuzzy logic concepts, by storing the query attributes and continuously refining the query to update the user's preferences. We have also built a Fuzzy Query system, which is a Java application that sits on top of a database.

With traditional SQL queries (relational DBMS), one can select records that match the selection criteria from a database. However, a record will not be selected if any one of the conditions fails. This makes searching for a range of potential candidates difficult.

In this program, one can basically retrieve all the records from the database, compare them with the desired record, aggregate the data, compute the ranking, and then output the records in the order of their rankings. Retrieving all the records from the database is a naïve approach because with some preprocessing, some very different records are not needed from the database. However, the main task is to compute the fuzzy rankings of the records so efficiency is not the main concern here.

The major difference between this application and other date matching system is that a user can input his hobbies in a fuzzy sense using a slider instead of choosing crisp terms like "kind of" or "love it". These values are stored in the database according to the slider value.

5.5 BISC-DSS: Applications

The followings are the potential applications of the BISC Decision Support System:

I. *Physical Stores or E-Store:* A computer system that could instantly track sales and inventory at all of its stores and recognize the customer buying trends and provide suggestion regarding any item that may interest the customer

- to arrange the products
- on pricing, promotions, coupons, etc
- for advertising strategy

II. *Profitable Customers:* A computer system that uses customer data that allows the company to recognize good and bad customer by the cost of doing business with them and the profits they return

- keep the good customers
- improve the bad customers or decide to drop them
- identify customers who spend money
- identify customers who are profitable

- compare the complex mix of marketing and servicing costs to access to new customers

III. *Internet-Based Advising: :* A computer system that uses the expert knowledge and the customer data (Internet brokers and full-service investment firms) to recognize the good and bad traders and provide intelligent recommendation to which stocks buy or sell

- reduce the expert needs at service centers
- increase customer confidence
- ease-of-use
- Intelligent coaching on investing through the Internet
- allow customers access to information more intelligently

IV. *Managing Global Business:* A computer system responding to new customers and markets through integrated decision support activities globally using global enterprise data warehouse

- information delivery in minutes
- lower inventories
- intelligent and faster inventory decisions in remote locations

V. *Resource Allocator:* A computer system that intelligently allocate resources given the degree of match between objectives and resources available

- resource allocation in factories floor
- for human resource management
- find resumes of applicants posted on the Web and sort them to match needed skill and can facilitate training and to manage fringe benefits programs
- evaluate candidates predict employee performance

VI. *Intelligent Systems to Support Sales:* A computer system that matching products and services to customers needs and interest based on case-based reasoning and decision support system to improve

- sale
- advertising

VII. *Enterprise Decision Support:* An interactive computer-based system that facilitates the solution of complex problems by a group of decision makers either by speeding up the process of the decision-making process and improving the quality of the resulting decisions through expert and user (company-customer) collaboration and sharing the information, goals, and objectives.

VIII. *Fraud Detection:* An Intelligent Computer that can learn the user's behavior through in mining customer databases and predicting customer behaviours (normal and irregularities) to be used to uncover, reduce or prevent fraud.

- in credit cards
- stocks
- financial markets
- telecommunication
- insurance

IX. *Supply-Chain Management (SCM):* Global optimization of design, manufacturing, supplier , distribution, planning decisions in a distributed environment

X. *BISC-DSS and Autonomous Multi-Agent System:* A key component of any autonomous multi-agent system –especially in an adversarial setting -- is decision module, which should be capable of functioning in an environment of imprecision, uncertainty and imperfect reliability. BISC-DSS will be focused on the development of such system and can be used as a decision-support system for ranking of decision alternatives. BISC-DSS can be used :

- As global optimizer for planning decisions in a distributed environment
- To facilitates the solution of complex problems by a group of autonomous agents by speeding up the process of decision-making, collaboration and sharing the information, goals, and objectives

- To intelligently allocate resources given the degree of match between objectives and resources available
- Assisting autonomous multi-agent system in assessing the consequences of decision made in an environment of imprecision, uncertainty, and partial truth and providing a systematic risk analysis
- Assisting multi-agent system answer “What if Questions”, examine numerous alternatives very quickly, ranking of decision alternatives, and find the value of the inputs to achieve a desired level of output

XI. *Knowledge Mobilization and Intelligent Augmentation (KnowMInA); Application to Homeland Security:* KnowMInA is an intelligent system that recognizes terrorism activities. ***KnowMInA*** includes behavior/profile modeling, reasoning engines, decision-risk analysis, and visual data mining-analytic software. KnowMInA is used to find suspicious pattern in data using geographical maps and recognition technology. It uses multimedia/Multimodal information or huge number of "tips" received prior or immediately after the attack. The system filters out the irrelevant part and acts on the useful subset of such data in a speedy manner. KnowMInA shares this relevant and summarized information with authorities using knowledge mobilization infrastructure. KnowMInA can detect unusual patterns; raise alarms based on classification of activities and offer explanations based on automatic learning techniques for why a certain activity is placed in a particular class such as “Safe”, “Suspicious”, “Dangerous” etc. The underlying techniques can combine expert knowledge and data driven rules to continually improve its classification and adapt to dynamic changes in data and expert knowledge. This will also provide the ability to answer “What if?” questions in order to decrease uncertainty and provide a better risk analysis. For Example, US authorities have announced that 5 out of 19 suspected September 11 hijackers met in Las Vegas before the attack. Some of the terrorist attended the same training school, had communication tracks, and flew at the same time period to the same location in the same flight. KnowMInA intends to capture and recognize such activities and raise alarms. KnowMInA would apply to the "Preparedness" and also "Response" part of the homeland security initiative.

6. Challenges and Road Ahead

World Wide Web search engines have become the most heavily-used online services, with millions of searches performed each day. Their popularity is due, in part, to their ease of use. The central tasks for the most of the search engines can be summarize as 1) query or user information request- do what I mean and not what I say!, 2) model for the Internet, Web representation-web page collection, documents, text, images, music, etc, and 3) ranking or matching function-degree of relevance, recall, precision, similarity, etc.

Design of any new intelligent search engine should be at least based on two main motivations:

i. The web environment is, for the most part, unstructured and imprecise. To deal with information in the web environment what is needed is a logic that supports modes of reasoning which are approximate rather than exact. While searches may retrieve thousands of hits, finding decision-relevant and query-relevant information in an imprecise environment is a challenging problem, which has to be addressed.

ii. Another, and less obvious, is deduction in an unstructured and imprecise environment given the huge stream of complex information.

Search engines, with Google at the top, have many remarkable capabilities. But what is not among them is the deduction capability—the capability to synthesize an answer to a query by drawing on bodies of information which are resident in various parts of the knowledge base. It is this capability that differentiates a question-answering system, Q/A system for short, from a search engine.

Prof. L. A. Zadeh in his recent work and several of his presentation has addressed the key issues related to importance of the use of Search Engine and the construction of Q/A system.

In this context, Berkeley Initiative in Soft Computing (BISC), University of California, Berkeley formed a Technical Committee to organize a Meeting entitled "Fuzzy Logic and the Internet: Enhancing the Power of the Internet" to understand the significance of the fields accomplishments, new developments and future directions. In addition, the Technical Committee selected and invited over 50 scientists (and industry experts as technical committee members) from the related disciplines to participate in the Meeting "State of the Art Assessment and New Directions for Research" which took place at the University of California, Berkeley, in August 14-18, 2001 and Nov 2003. The Workshops provided a unique opportunity for the academic and corporate communities to address new challenges, share solutions, and discuss research directions for the future. Following are the areas that were recognized as challenging problems and the new direction toward the next generation of the search engines and Internet. We summarize the challenges and the road ahead into four categories as follows:

Search Engine and Queries:	***Internet and the Academia:***
• *Deductive Capabilities* • *Customization and Specialization* • *Metadata and Profiling* • *Semantic Web* • *Imprecise-Querying* • *Automatic Parallelism via Database Technology* • *Approximate Reasoning* • *Ontology* • *Ambiguity Resolution through Clarification Dialog; Definition/Meaning & SpecificityUser Friendly* • *Multimedia* • *Databases* • *Interaction*	• *Ambiguity and Conceptual and Ontology* • *Aggregation and Imprecision Query* • *Meaning and structure Understanding* • *Dynamic Knowledge* • *Perception, Emotion, and Intelligent Behavior* • *Content-Based* • *Escape from Vector SpaceDeductive Capabilities* • *Imprecise-Querying* • *Ambiguity Resolution through Clarification Dialog* • *Precisiated Natural Languages (PNL)*

Internet and the Industry:	***Fuzzy Logic and Internet; Fundamental Research:***
• *XML=>Semantic Web* • *Workflow* • *Mobile E-Commerce* • *CRM* • *Resource Allocation* • *Intent* • *Ambiguity Resolution* • *Interaction* • *Reliability* • *Monitoring* • *Personalization and Navigation* • *Decision Support* • *Document Soul* • *Approximate Reasoning* • *Imprecise QueryContextual Categorization*	• *Computing with Words (CW)* [Zadeh, (1996,), (1999); Zadeh, Kacprzyk, (1999a) and (1999b)]. • *Computational Theory of Perception (CTP)* [Zadeh, (2001b); Zadeh, Nikravesh, (2002)]. • *Precisiated Natural Languages (PNL)*

The potential areas and applications of Fuzzy Logic for the Internet include:

- ***Potential Areas:***
  - *Search Engines*
  - *Retrieving Information*
  - *Database Querying*
  - *Ontology*
  - *Content Management*
  - *Recognition Technology*
  - *Data Mining*
  - *Summarization*
  - *Information Aggregation and Fusion*
  - *E-Commerce*
  - *Intelligent Agents*
  - *Customization and Personalization*

- ***Potential Applications:***
  - *Search Engines and Web Crawlers*
  - *Agent Technology (i.e., Web-Based Collaborative and Distributed Agents)*
  - *Adaptive and Evolutionary techniques for dynamic environment (i.e. Evolutionary search engine and text retrieval, Dynamic learning and adaptation of the Web Databases, etc)*
  - *Fuzzy Queries in Multimedia Database Systems*
  - *Query Based on User Profile*
  - *Information Retrievals*
  - *Summary of Documents*

- *Information Fusion Such as Medical Records, Research Papers, News, etc*
- *Files and Folder Organizer*
- *Data Management for Mobile Applications and eBusiness Mobile Solutions over the Web*
- *Matching People, Interests, Products, etc*
- *Association Rule Mining for Terms-Documents and Text Mining*
- *E-mail Notification*
- *Web-Based Calendar Manager*
- *Web-Based Telephony*
- *Web-Based Call Centre*
- *Workgroup Messages*
- *E-Mail and Web-Mail*
- *Web-Based Personal Info*
- *Internet related issues such as Information overload and load balancing, Wireless Internet-coding and D-coding (Encryption), Security such as Web security and Wireless/Embedded Web Security, Web-based Fraud detection and prediction, Recognition, issues related to E-commerce and E-bussiness, etc.*

7. Conclusions

Intelligent search engines with growing complexity and technological challenges are currently being developed. This requires new technology in terms of understanding, development, engineering design and visualization. While the technological expertise of each component becomes increasingly complex, there is a need for better integration of each component into a global model adequately capturing the imprecision and deduction capabilities. In addition, intelligent models can mine the Internet to conceptually match and rank homepages based on predefined linguistic formulations and rules defined by experts or based on a set of known homepages. The FCM model can be used as a framework for intelligent information and knowledge retrieval through conceptual matching of both text and images (here defined as "Concept"). The FCM can also be used for constructing fuzzy ontology or terms related to the context of the query and search to resolve the ambiguity. This model can be used to calculate conceptually the degree of match to the object or query.

It is important to note that while the Semantic Web is dissimilar in many ways from the World Wide Web and search Engine, the **Semantic Web** provides a common framework that allows **data** to be shared and reused across application, enterprise, and community boundaries through the World Wide Web. In this paper, we presented a framework that goes beyond the traditional semantic web which has been defined mostly as a mesh or distributed databases within the World Wide Web. In this paper, we focused on development of a framework for reasoning and deduction in the web. A web-based model to decision model for analysis of structured database has been presented. In addition, a framework to incorporate the information from web sites into the search engine has been presented as a model that will go beyond current semantic web idea.

Acknowledgements

Funding for this research was provided by the British Telecommunication (BT) and the BISC Program of UC Berkeley.

References

J. Baldwin, Future directions for fuzzy theory with applications to intelligent agents, in M. Nikravesh and B. Azvine, FLINT 2001, New Directions in Enhancing the Power of the Internet, UC Berkeley Electronics Research Laboratory, Memorandum No. UCB/ERL M01/28, August 200.

J. F. Baldwin and S. K. Morton, conceptual Graphs and Fuzzy Qualifiers in Natural Languages Interfaces, 1985, University of Bristol.

M. J. M. Batista et al., User Profiles and Fuzzy Logic in Web Retrieval, in M. Nikravesh and B. Azvine, FLINT 2001, New Directions in Enhancing the Power of the Internet, UC Berkeley Electronics Research Laboratory, Memorandum No. UCB/ERL M01/28, August 2001.

H. Beremji, Fuzzy Reinforcement Learning and the Internet with Applications in Power Management or wireless Networks, in M. Nikravesh and B. Azvine, FLINT 2001, New Directions in Enhancing the Power of the Internet, UC Berkeley Electronics Research Laboratory, Memorandum No. UCB/ERL M01/28, August 2001.

T.H. Cao, Fuzzy Conceptual Graphs for the Semantic Web, in M. Nikravesh and B. Azvine, FLINT 2001, New Directions in Enhancing the Power of the Internet, UC Berkeley Electronics Research Laboratory, Memorandum No. UCB/ERL M01/28, August 2001.

D. Y. Choi, Integration of Document Index with Perception Index and Its Application to Fuzzy Query on the Internet, in M. Nikravesh and B. Azvine, FLINT 2001, New Directions in Enhancing the Power of the Internet, UC Berkeley Electronics Research Laboratory, Memorandum No. UCB/ERL M01/28, August 2001.

R. Fagin (1998) Fuzzy Queries in Multimedia Database Systems, Proc. ACM Symposium on Principles of Database Systems, pp. 1-10.

R. Fagin (1999) Combining fuzzy information from multiple systems. J. Computer and System Sciences 58, pp 83-99.

K.H.L. Ho, Learning Fuzzy Concepts by Example with Fuzzy Conceptual Graphs. In 1st Australian Conceptual Structures Workshop, 1994. Armidale, Australia.

J. H. Holland. Adaptation in Natural and Artificial Systems: An Introductory Analysis with Applications to Biology, Control and Artificial Intelligence. MIT Press, 1992. First Published by University of Michigan Press 1975

A. Joshi and R. Krishnapuram, Robust Fuzzy Clustering Methods to Support Web Mining, in Proc Workshop in Data Mining and Knowledge Discovery, SIGMOD, pp. 15-1 to 15-8, 1998.

M. Kobayashi, K. Takeda, "Information retrieval on the web", ACM Computing Survey, Vol.32, pp.144-173 (2000) \

J. R. Koza, Genetic Programming : On the Programming of Computers by Means of Natural Selection, Cambridge, Mass. : MIT Press, USA 1992, 819 pages

R. Krishnapuram et al., A Fuzzy Relative of the K-medoids Algorithm with application to document and Snippet Clustering , in Proceedings of IEEE Intel. Conf. Fuzzy Systems-FUZZIEEE 99, Korea, 1999.

V. Loia et al., "Fuzzy Logic an the Internet", in the Series Studies in Fuzziness and Soft Computing, Physica-Verlag, Springer (August 2003)

V. Loia et al., *Journal of Soft Computing*, Special Issue; fuzzy Logic and the Internet, Springer Verlag, Vol. 6, No. 5; August 2002.

T. P. Martin, Searching and smushing on the Semantic Web – Challenges for Soft Computing, in M. Nikravesh and B. Azvine, FLINT 2001, New Directions in Enhancing the Power of the Internet, UC Berkeley Electronics Research Laboratory, Memorandum No. UCB/ERL M01/28, August 2001.

M. Mizumoto (1989) Pictorial Representations of Fuzzy Connectives, Part I: Cases of T-norms, T-conorms and Averaging Operators, Fuzzy Sets and Systems 31, pp. 217-242.

M. Nikravesh (2001a) Perception-based information processing and retrieval: application to user profiling, 2001 research summary, EECS, ERL, University of California, Berkeley, BT-BISC Project. (http://zadeh.cs.berkeley.edu/ & http://www.cs.berkeley.edu/~nikraves/ & http://www-bisc.cs.berkeley.edu/).

M. Nikravesh (2001b), BISC and The New Millennium, Perception-based Information Processing, Berkeley Initiative in Soft Computing, Report No. 2001-1-SI, September 2001b.

M. Nikravesh (2001c) Credit Scoring for Billions of Financing Deci-sions, Joint 9th IFSA World Congress and 20th NAFIPS International Conference. IFSA/NAFIPS 2001 "Fuzziness and Soft Computing in the New Millenium", Vancouver, Canada, July 25-28, 2001.

M. Nikravesh, B. Azvine, R. Yagar, and Lotfi A. Zadeh (2003) "New Directions in Enhancing the power of the Internet", Editors:, in the Series Studies in Fuzziness and Soft Computing, Physica-Verlag, Springer (August 2003)

M. Nikravesh and B. Azvine, FLINT 2001, New Directions in Enhancing the Power of the Internet, UC Berkeley Electronics Research Laboratory, Memorandum No. UCB/ERL M01/28, August 2001.

M. Nikravesh et al., "Enhancing the Power of the Internet", in the Series Studies in Fuzziness and Soft Computing, Physica-Verlag, Springer (August 2003) (2003a)

M. Nikravesh, et al., Perception-Based Decision processing and Analysis, UC Berkeley Electronics Research Laboratory, Memorandum No. UCB/ERL M03/21, June 2003 (2003b).

M. Nikravesh and D-Y. Choi, Perception-Based Information Processing, UC Berkeley Electronics Research Laboratory, Memorandum No. UCB/ERL M03/20, June 2003.

M. Nikravesh et al., Web Intelligence: Conceptual-Based Model, UC Berkeley Electronics Research Laboratory, Memorandum No. UCB/ERL M03/19, June 2003 (2003c).

M. Nikravesh et al., Fuzzy logic and the Internet (FLINT), Internet, World Wide Web, and Search Engines, *Journal of Soft Computing*, Special Issue; fuzzy Logic and the Internet, Springer Verlag, Vol. 6, No. 5; August 2002

M. Nikravesh, Fuzzy Conceptual-Based Search Engine using Conceptual Semantic Indexing, NAFIPS-FLINT 2002, June 27-29, New Orleans, LA, USA

M. Nikravesh and B. Azvine, Fuzzy Queries, Search, and Decision Support System, *Journal of Soft Computing*, Special Issue fuzzy Logic and the Internet, Springer Verlag, Vol. 6, No. 5; August 2002.

S. K. Pal, V. Talwar, and P. Mitra, Web Mining in Soft Computing Framework: Relevance, State of the Art and Future Directions, to be published in IEEE Transcations on Neural Networks, 2002.

G. Presser, Fuzzy Personalization, in M. Nikravesh and B. Azvine, FLINT 2001, New Directions in Enhancing the Power of the Internet, UC Berkeley Electronics Research Laboratory, Memorandum No. UCB/ERL M01/28, August 2001.

E. Sanchez, Fuzzy logic e-motion, in M. Nikravesh and B. Azvine, FLINT 2001, New Directions in Enhancing the Power of the Internet, UC Berkeley Electronics Research Laboratory, Memorandum No. UCB/ERL M01/28, August 2001.

A. M. G. Serrano, Dialogue-based Approach to Intelligent Assistance on the Web, in M. Nikravesh and B. Azvine, FLINT 2001, New Directions in Enhancing the Power of the Internet, UC Berkeley Electronics Research Laboratory, Memorandum No. UCB/ERL M01/28, August 2001.

S. Shahrestani, Fuzzy Logic and Network Intrusion Detection, in M. Nikravesh and B. Azvine, FLINT 2001, New Directions in Enhancing the Power of the Internet, UC Berkeley Electronics Research Laboratory, Memorandum No. UCB/ERL M01/28, August 2001.

T. Takagi, A. Imura, H. Ushida, and T. Yamaguchi, “Conceptual Fuzzy Sets as a Meaning Representation and their Inductive Construction,” International Journal of Intelligent Systems, Vol. 10, 929-945 (1995).

T. Takagi, A. Imura, H. Ushida, and T. Yamaguchi, “Multilayered Reasoning by Means of Conceptual Fuzzy Sets,” International Journal of Intelligent Systems, Vol. 11, 97-111 (1996).

T. Takagi, S. Kasuya, M. Mukaidono, T. Yamaguchi, and T. Kokubo, “Realization of Sound-scape Agent by the Fusion of Conceptual Fuzzy Sets and Ontology,” 8th International Conference on Fuzzy Systems FUZZ-IEEE'99, II, 801-806 (1999).

T. Takagi, S. Kasuya, M. Mukaidono, and T. Yamaguchi, "Conceptual Matching and its Applications to Selection of TV Programs and BGMs," IEEE International Conference on Systems, Man, and Cybernetics SMC'99, III, 269-273 (1999).

T. Takagi, et al., Conceptual Fuzzy Sets as a Meaning Representation and their Inductive Construction, International Journal of Intelligent Systems, Vol. 10, 929-945 (1995).

T. Takagi and M.Tajima, Proposal of a Search Engine based on Conceptual Matching of Text Notes, IEEE International Conference on Fuzzy Systems FUZZ-IEEE'2001, S406- (2001a)

T. Takagi and M. Tajima, Proposal of a Search Engine based on Conceptual Matching of Text Notes, in M. Nikravesh and B. Azvine, FLINT 2001, New Directions in Enhancing the Power of the Internet, UC Berkeley Electronics Research Laboratory, Memorandum No. UCB/ERL M01/28, August 2001 (2001b).

T. Takagi, et al., Exposure of Illegal Website using Conceptual Fuzzy Sets based Information Filtering System, the North American Fuzzy Information Processing Society - The Special Interest Group on Fuzzy Logic and the Internet NAFIPS-FLINT 2002, 327-332 (2002a)

T. Takagi, et al., Conceptual Fuzzy Sets-Based Menu Navigation System for Yahoo!, the North American Fuzzy Information Processing Society - The Special Interest Group on Fuzzy Logic and the Internet NAFIPS-FLINT 2002, 274-279 (2002b)

Wittgenstein, "Philosophical Investigations," Basil Blackwell, Oxford (1953).

R. Yager, Aggregation Methods for Intelligent Search and Information Fusion, in M. Nikravesh and B. Azvine, FLINT 2001, New Directions in Enhancing the Power of the Internet, UC Berkeley Electronics Research Laboratory, Memorandum No. UCB/ERL M01/28, August 2001.

J. Yen, Incorporating Fuzzy Ontology of Terms Relations in a Search Engine, in M. Nikravesh and B. Azvine, FLINT 2001, New Directions in Enhancing the Power of the Internet, UC Berkeley Electronics Research Laboratory, Memorandum No. UCB/ERL M01/28, August 2001.

L. A. Zadeh, Fuzzy Logic = Computing with Words," *IEEE Trans. on Fuzzy Systems* (4), 103-111, 1996.

L. A. Zadeh, (1999) From Computing with Numbers to Computing with Words-From Manipulation of Measurements to Manipulation of Perceptions, *IEEE Trans. On Circuit and Systems-I Fundamental Theory and Applications,* 45(1), Jan 1999, 105-119.

L. A. Zadeh, "A new direction in AI – Toward a computational theory of perceptions", *AI Magazine* 22(1): Spring 2001, 73-84, 2001b.

L. A. Zadeh, and J. Kacprzyk, (eds.), Computing With Words in Information/Intelligent Systems 1: Foundations, Physica-Verlag, Germany, 1999a.

L. A. Zadeh, L. and J. Kacprzyk (eds.), Computing With Words in Information/Intelligent Systems 2: Applications, Physica-Verlag, Germany, 1999b.

L. A. Zadeh, The problem of deduction in an environment of imprecision, uncertainty, and partial truth, in M. Nikravesh and B. Azvine, FLINT 2001, New Directions in Enhancing the Power of the Internet, UC Berkeley Electronics Research Laboratory, Memorandum No. UCB/ERL M01/28, August 2001 (2001).

L. A. Zadeh, A Prototype-Centered Approach to Adding Deduction Capability to Search Engines -- The Concept of Protoform, BISC Seminar, Feb 7, 2002, UC Berkeley, 2002.

L.A. Zadeh, Toward a Perception-based Theory of Probabilistic Reasoning with Imprecise Probabilities, *Journal of Statistical Planning and Inference*, 105 233–264, 2002.

L. A. Zadeh and M. Nikravesh, Perception-Based Intelligent Decision Systems; Office of Naval Research, Summer 2002 Program Review, Covel Commons, University of California, Los Angeles, July 30th-August 1st, 2002.

Y. Zhang et al., Granular Fuzzy Web Search Agents, in M. Nikravesh and B. Azvine, FLINT 2001, New Directions in Enhancing the Power of the Internet, UC Berkeley Electronics Research Laboratory, Memorandum No. UCB/ERL M01/28, August 2001 (2001a).

Y. Zhang et al., Fuzzy Neural Web Agents for Stock Prediction, in M. Nikravesh and B. Azvine, FLINT 2001, New Directions in Enhancing the Power of the Internet, UC Berkeley Electronics Research Laboratory, Memorandum No. UCB/ERL M01/28, August 2001 (2001b).

An Ontology-Based Method for User Model Acquisition

Célia da Costa Pereira[1] and Andrea G. B. Tettamanzi[1]

Università Degli Studi di Milano
Dipartimento di Tecnologie dell'Informazione
Via Bramante 65, I-26013 Crema (CR), Italy
pereira@dti.unimi.it, andrea.tettamanzi@unimi.it

Summary. This chapter illustrates a novel approach to learning user interests from the way users interact with a document management system. The approach is based on a fuzzy conceptual representation of both documents and user interests, using information contained in an ontology. User models are constructed and updated by means of an on-line evolutionary algorithm. The approach has been successfully implemented in a module of a knowledge management system which helps the system improve the relevance of responses to user queries.

1 An Introduction to User Model Acquisition

Research in User Model Acquisition (UMA) has received considerable attention in the last years. As the amount of information available online grows with astonishing speed, people feel overwhelmed navigating through today's information and media landscape. The problem of information overload leads to a demand for automated methods to locate and retrieve information with respect to users' individual interests. Unfortunately, methods developed within information retrieval leave two main problems still open.

The first problem is that most approaches assume users to have a well-defined idea of what they are looking for, which is not always the case. We solve this problem by letting fuzzy user models evolve on the basis of a rating induced by user behavior.

The second problem concerns the use of *keywords*, not *concepts*, to formulate queries. Considering words and not the concepts behind them often leads to a loss in terms of the quantity and quality of information retrieved. We solve this problem by adopting an ontology-based approach.

The approach described in this paper has been implemented and successfully tested in the framework of the Information and Knowledge Fusion (IKF) Eureka Project E!2235 [8]. The motivation of this work, was to propose an adaptive method for acquiring user models for the users of the IKF

C. da C. Pereira and A.G.B. Tettamanzi: *An Ontology-Based Method for User Model Acquisition*, StudFuzz **204**, 211–229 (2006)
www.springerlink.com

(Information Knowledge Fusion) system. That system may be regarded as a knowledge-aware document management system offering advanced search capabilities to its users. However, a more familiar, but entirely appropriate, scenario for understanding the approach may be a document retrieval problem like the ones solved by WWW search engines.

The idea is to develop a system that knows its users. If a user has already interacted with the system, the system is capable of addressing her directly to the feature she is interested in. In order to make that possible, the first time a user logs in, the system constructs an initial model of the user. Then, for each of the user's subsequent sessions, the existing model is updated by creating or adding new information or deleting information which has become invalid. An accurate and reliable model would allow for quick and efficient answers to user demands.

In Sect. 2, we survey some of the existing works in the UMA field, those with common characteristics with the approach we propose here. In Sect. 3 we carry out a general presentation as regards our approach to the UMA; Sect. 4 describes a genetic algorithm for evolving user models, and Sect. 5 provides a discussion of the approach.

2 Related Work on User Model Acquisition

Techniques for User Model Acquisition are used in different fields. In this section we present some of the contexts in which techniques for UMA have been devised in recent years.

2.1 UMA for the Worldwide Web

In the Worldwide Web (WWW) framework, the main goal of UMA is to "learn" user interests from information obtained each time she/he logs in and chooses a page to read. In [27] the authors propose a novel method for learning models of the user's browsing behavior from a set of annotated web logs. A similar work is described in [5], in which user interest profiles are automatically costructed by monitoring user web and email habits. A clustering algorithm is employed to identify interests, which are then clustered together to form interest themes. In Jasper [7] the authors have costructed a distributed system of intelligent agents for performing information tasks over the WWW on behalf of a community of users. Jasper can summarize and extract keywords from WWW pages and can share information among users with similar interests automatically. Experimental work has been carried out to investigate user profiling within a framework for personal agents [21]. The approach is based around the electronic documents a user reads or writes. Typically, when a user marks a page as interesting (or uninteresting), or the user visits a particular page, an agent performs some analysis of that page in order to determine what feature(s) caused the user to be interested (or uninterested) in it. More recently, in [27] a novel method is presented for learning models of the user's browsing behavior from a set of annotated web logs — i.e., web logs that are augmented with the user's assessment of whether each webpage is an *information content* (IC) page (i.e., it contains the information required to complete her task). The system uses this kind of information to learn what properties of a webpage, within a sequence, identify such IC-pages. According to the authors, as these methods deal with *properties* of webpages (or words), rather than specific URLs, they can be used anywhere throughout the web; i.e., they are not specific to a particular website, or a particular task.

2.2 UMA for Recommendation Systems

Another field in which UMA techniques are applied is recommendation systems. A recommendation system tracks past actions of a group of users to make recommendations to individual members of the group. MORSE (movie recommendation system) [10] makes personalized movie recommendations based on what is known about users' movie preferences. These informations are provided to the system by users ratings about movies they have seen on a numeric scale. MORSE is based on the principle of social filtering. The accuracy of its recommendations improves as more people use the system and as more movies are rated by individual users. In [19], a content-based recommendation component is developed for an information system named ELFI. This system learns individual interest profiles from observation considering positive evidence only. In the case in which standard classification methods are not appropriate, the authors use both a probabilistic and an instance-based approach to characterize the set of information objects a user observed to interact with in a positive way. Another approach based on collaborative information filtering [4] consists in predicting items a user would like on the basis of other user's ratings for these items. The best-performing algorithm proposed in this work is based on the singular value decomposition of an initial matrix of user ratings, exploiting latent structure that essentially eliminates the need for users to rate common items in order to become predictors for one another's preferences.

2.3 UMA for Information Retrieval

The idea behind UMA is also used in information rietrieval (IR) approaches. In [11], the authors propose a system that carries out highly effective searches over collections of textual information, such as those found on the Internet. The system is comprised of two major parts. The first part consists of an agent, MUSAG, that learns to relate concepts that are semantically "similar" to one another. In other words, this agent dynamically builds a dictionary of expressions for a given concept. The second part consists of another agent, SAg, which is responsible for retrieving documents, given a set of keywords with relative weights. This retrieval makes use of the dictionary learned by

MUSAG, in the sense that the documents to be retrieved for a query are semantically related to the concept given. A related approach is proposed in [17]: an IR method is described in which what is considered is the semantic sense of a word, not the word itself. The authors have constructed an agent for a bilingual news web site — SiteF —, which learns user interests from the pages the users request. The approach uses a content-based document representation as a starting point to build a model of the user's interests. As the user browses the documents, the system builds the user model as a semantic network whose nodes represent senses (as opposed to keywords) that identify the documents requested by the user. A filtering procedure dynamically predicts new documents on the basis of that semantic network.

2.4 UMA for Predictive Systems

UMA techniques are also used in predictive systems, as it is the case with the LISTEN Project [2]. That project comprises an "intelligent tutor" for reading, which constructs and uses a fine-grained model of students. Model construction uses a database describing student interactions with their tutor to train a classifier that predicts whether students will click on a particular word for help with 83% accurancy. The goal of the authors is to strengthen the Reading Tutor's user model and determine which user characteristics are predictive of student behavior. Thus doing, if the tutor knows when the student will, for example, require assistance, it can provide help in a proactive way.

A recent overview on how to apply machine learning techniques to acquire and continuously adapt user models [24] describes a novel representation of user models as Bayesian networks (BNs). The generic framework considered by the autor is flexible with regard to several points such as:

- *off-line learning* and *on-line adaptation*. During the off-line phase, the general User Model is learned on the basis of data from other previous system users or data acquired by user studies. These models are in turn used as a starting point for the interaction with a particular new user. The initial general model is then adapted to the individual current user and can be saved after the interaction for future use when this particular user will interact the next time with the system.
- *Experimental data and usage data.* Experimental data, which mostly does not represent the "reality", is collected in controlled environments just as done in psychological experiments. Usage data, which often includes more missing data and rare situations, are underrepresented in such data sets; such data are collected during the real interaction between users and the system.
- *Learning the BN conditional probabilities and structures.* The learning and adaptation tasks are two-dimensional: learning the conditional probabilities and learning the BN structures. To deal with sparse data the autor has introduced additional available domain knowledge into the learning procedures to improve the results, especially when the data is indeed sparse.

- *Degree of interpretability*: the author tries to ensure or at least improve the interpretability of the final learned models, e.g., by respecting whatever background information that is *a priori* available and that can be introduced into and exploited within the learning process.

3 An Ontology-Based Approach

We assume concepts and their relations to be organized in a formal ontology [13, 22]. It is convenient to logically divide an ontology into three levels[1]: the *top-level* ontology, containing the most general concepts, valid for all applications; a *middle-level* ontology, comprising more specific concepts which are shared by several applications in the same broad area (e.g., Finance, Manufacturing); a *domain* ontology, comprising all the technical concepts relevant to a specific domain or sub-domain (e.g., Banking, Automotive Industry, Internet Services).

The key advantage an ontological approach offers is that one can work with *concepts* rather than with *keywords*: we will assume documents have already been processed by a suitable natural language processing technique (for instance, a semantic parser) and the appropriate sense (i.e., concept) has been associated to each word or phrase in the document[2].

For example, when a user is looking for a place to stay, the system should automatically consider all places in which it is possible to stay more than one day, i.e., houses, apartments, bed-and-breakfasts, hotels, etc., without requiring the user to enter all these (and possibly other) keywords. Precisely, the system will consider all the concepts subsumed by the concept "place to stay" according to the ontology. This consideration allows us to solve a common problem existing in many UMA approaches, concerning the loss of information due to the simple use of keywords instead of the concepts they stand for.

We have adopted a collaborative filtering [4, 10] approach for rating users preferences. Consequently, if a new user has a profile similar to a set of users already existing in the system, a set of documents interesting for these users is also proposed to the new user. The central idea of collaborative filtering is to base personalized recommendations for users on information obtained from other, ideally likemind, users.

[1] Of course, it is possible to consider the division on fewer or more levels dipending on the problem one is solving

[2] The knowledge management system for which the UMA approach here described has been developed actually uses a state-of-the art semantic parser, which takes advantage of a lexical database derived from WordNet [9], and an array of pattern- and rule-based transformations to semantically annotate documents.

3.1 Document Representation

In this framework, a document is represented as a vector, whose components express the "importance" of a concept, instead of a keyword, as it will be explained in this section.

Passing from an unstructured set of (key)words seen as independent entities to an ontology of concepts, structured by the hierarchical *is-a* relation into a lattice, is not trivial. Suppose, for example, that we find in a document the three related concepts of *cat*, *dog*, and *animal*: now, *animal* is a super-class of both *cat* and *dog*. Therefore, mentioning the *animal* concept can implicitly refer both to *cat* and *dog*, although not so specifically as mentioning *cat* or *dog* directly would. Therefore, we need to devise a system to take this kind of interdependence among concepts into account when calculating levels of importance.

Choice of the Concepts to Represent a Document

The main idea is to consider but the leaf concepts in the ontology, i.e., the most specific concepts only, as elements of the importance vector for a document. This choice can be justified by thinking that more general concepts (i.e., internal concepts) are implicitly taken account of through the leaf concepts they subsume. As a matter of fact, considering internal concepts too, along with their sub-classes, would create overlaps or dependencies in the importance vector, in that the component measuring the importance of an internal concept should be a function of all the components measuring the importance of all of its sub-classes, down to the leaf concepts subsumed by it.

Instead, by considering the leaf concepts only, the concepts in the document that are internal nodes of the ontology lattice are implicitly represented in the importance vector by "distributing" their importance to all of their sub-classes down to the leaf concepts in equal proportion. By doing so, we are sure that our document vector is expressed relative to concepts such that one does not subsume another. In other words, the components of the vector are independent, or orthogonal, with respect to each other.

Since an ontology (consisting of all the three layers mentioned above) can potentially contain many concepts (e.g., in the order of the hundreds of thousands), it is convenient to limit ourselves, for the purpose of calculating document importance vectors, only to the top-level ontology, whose size may be in the thousands of concepts. Anyway, it is important to remark that the approach described below is absolutely independent of which subset of the ontology one decides to restrict to, and that what we call "leaf concepts" can be defined in any way that is consistent with our premise, that is, that they form a set of mutually independent concepts (i.e., none of them subsumes, or is subsumed by, any other). The ontology lattice, cut this way, can be treated as a direct acyclic graph (*dag*), having $\top$ (the most general concept) as its root.

Representing a Document as a Vector of Concepts

The quantity of information given by the presence of some concept c in a document depends on the depth of c in the ontology graph, on how many times it appears in the document, and how many times it occurs in the whole information repository. These two frequencies also depend on the number of concepts which subsume c. Let us consider a concept a which is a descendant of another concept b which has q children including a. Concept b is a descendant of a concept c which has k children including b. Concept a is a leaf of the graph representing our ontology. If we consider a document containing only "ab", the occurrence of a in the document is $1+\frac{1}{q}$. In the document "abc" the occurrence of a is $1+\frac{1}{q}(1+\frac{1}{k})$. As we can see, the number of occurrences of a leaf is proportional to the number of children which all of its ancestors have.

Explicit and implicit concepts are taken into account by using the following formulas:

$$N(c) = \mathrm{occ}(c) + \sum_{\mathtt{c}\in\mathtt{Path}(c,\ldots,\top)} \sum_{i=2}^{\mathtt{length}(\mathbf{c})} \frac{\mathrm{occ}(c_i)}{\prod_{j=2}^{i} \|\mathrm{children}(c_j)\|}, \tag{1}$$

where $N(c)$ is the count of explicit and implicit occurrences of concept c, and $\mathrm{occ}(c)$ is the number of occurrences of lexicalizations of c.

The quantity of information $\mathrm{info}(c)$ given by the presence of some concept c in a document is given by

$$\mathrm{info}(c) = \frac{N_{doc}(c)}{N_{rep}(c)}, \tag{2}$$

where $N_{doc}(c)$ is the number of times a lexicalization of c appears in the document, and $N_{rep}(c)$ is the total number of its (explicit as well as implicit) occurrences in the whole document repository.

Each document is then represented by a document vector I, which can be understood as a concept importance vector, whose jth component is

$$I(c_j) = \frac{\mathrm{info}(c_j)}{\sum_{i\in\mathcal{L}} \mathrm{info}(i)}, \tag{3}$$

where c_j is a leaf concept, and $\mathcal{L}$ is the set of all leaf concepts.

We define similarity between documents on the basis of their vector representation, and we group them into clusters according with their similarity, in order to express user interests with respect to clusters instead of all individual documents.

3.2 Grouping Documents by Interest

The model of a user of the system contains her/his preferences about the documents of the repository. To optimize the space required to represent them, we decide to group all documents with a certain level of similarity and associate a label to that group.

Similarity between Two Documents

To compare two document vectors, we consider their shared and exclusive concepts with their respective importance degrees. We suppose that the presence of a very specific concept in a document and its absence in the other document increases the difference between them. Two functions to calculate the similarity between two documents d_i and d_j are proposed here. The first one is based on the Euclidean distance between two vectors. It is given by:

$$\mathrm{sim}(d_i, d_j) = e^{-\lambda D(d_i, d_j)} \tag{4}$$

where $D(d_i, d_j)$ is the Euclidean distance between document vectors d_i and d_j and λ is a non negative coefficient which represents how quickly the distance between two document vectors influences their similarity.

Let I_i and I_j be the concept importance vectors corresponding to the documents d_i and d_j respectively. $D(d_i, d_j)$ is then computed as follows:

$$D(d_i, d_j) = \sqrt{\sum_k^K |I_i(c_k) - I_j(c_k)|^2} \tag{5}$$

where K is the number of leaf concepts in the ontology graph.

The second proposition is based on both the importance of concepts co-occurring in the documents and on the concepts which are present in a document and not in the other. It is given by:

$$\mathrm{sim}(d_i, d_j) = \frac{|d_i \wedge d_j|}{|d_i \wedge d_j| + \alpha(d_i, d_j) + \alpha(d_j, d_i)} \tag{6}$$

where

$$|d_i \wedge d_j| = \sum_{c \models d_i \wedge d_j} \min(I_i(c), I_j(c)). \tag{7}$$

which quantifies the importance of the number of the concepts co-occurring in documents d_i and d_j.

In opposition, the function α allows us to quantify the importance of concepts which are exclusive to one or the other document. It is defined as:

$$\alpha(d_i, d_j) = \sum_{c \models d_i \wedge \bar{d}_j} I_i(c). \tag{8}$$

Grouping Similar Documents into Fuzzy Clusters

Due to the availability of ever larger numbers of text documents in digital form and to the ensuing need to organize them for easier use, the dominant approaches to classify documents is one of building text classifiers automatically. Approaches based on machine learning have been proposed [12, 25, 20].

Many other approaches are based on clustering algorithms [26, 23]. Recently an approach has been proposed [16], which is based on a newly developped algorithm learning very large tree-like Bayesian networks.

In this work we use a clustering-based method [14, 15] to classify documents. By adopting a clustering method for grouping documents, we aim at keeping together all documents which are similar. By doing so, we have the following advantages:

- the number of clusters is much less than the number of documents. This fact may allow us to reduce considerably the computational cost of our evolutionary algorithm and thus to accelerate the search time;
- because similar documents are grouped in the same cluster, if a document is interesting to a user, the documents in the same cluster have a big chance to be interesting too to the user. This fact presents a particular advantage in the sense that it dispenses us from making a number of comparisons which is proportional to the size of the cluster.

It is likely that a given document may fit, to some extent, into more than one cluster. Any fuzzy clustering algorithm, like fuzzy c means [3], can serve this purpose.

The fuzzy clustering algorithm we have chosen for grouping the documents is based on the similarity of the documents. The three main objectives of the algorithm are:

- maximize the similarity sim between the documents in the same cluster,
- maximize the distance between the centers of the clusters,
- minimize the number of clusters.

A document in a fuzzy cluster resulting from our algorithm may also represent more than one interest and thus belong to more than one cluster.

A fuzzy cluster i is then represented as:

$$i = \{(d_{i1}, \mu_i(d_{i1})), \ldots, (d_{in}, \mu_i(d_{in}))\}, \tag{9}$$

where n is the number of documents in the cluster i, d_{ij} is the jth document in the cluster and $\mu_i(d_{ij})$ is the degree with which document d_{ij} belongs to the cluster i (its membership degree).

If K is the number of clusters, for all document d,

$$\sum_{k=1}^{K} \mu_k(d) = 1, \tag{10}$$

which represents the fact that, while one document may belong to more than one cluster, the clusters form a fuzzy partition of the document space.

No cluster is empty and and no cluster is the whole set of documents in the information repository. Thus, for all cluster i:

$$0 < \sum_{k=1}^{n} \mu_i(d_{ik}) < n. \tag{11}$$

Each cluster i has a *cluster center* v_i such that the importance of the jth coordinate $v_i(c_j)$ is calculated as:

$$v_i(c_j) = \frac{\sum_{k=1}^{n} \mu_i^m(d_{ik}) \cdot I_k(c_j)}{\sum_{k=1}^{n} \mu_i^m(d_{ik})} \tag{12}$$

where $m \geq 1$ is the weighting parameter controlling the amount of fuzziness of the clustering process.

The clustering algorithm is used during the initial phase of the acquisition process, and at any time it will be necessary. That's because, initially the information repository contains a certain number of documents, but it is possible that new documents are inserted (or existing documents are removed) during the acquisition process. For this reason, we decided to update the contents of the repository every T periods. The value of T depends on how many new documents have been inserted (or how many existing documents have been removed).

In our approach, we consider that all the documents in the same cluster must have a similarity degree greater than a threshold ϑ. The general principle of the fuzzy clustering algorithm is:

1. consider two documents d_i and d_j such that $\mathrm{sim}(d_i, d_j) < \vartheta$. Construct then two different clusters $C_i = \{d_i\}$ and $C_j = \{d_j\}$.
2. for each other documents d_k out of any cluster compute the similarity with the center v_i of the existing clusters C_i: $\mathrm{sim}(d_k, v_i)$. If all these similarities are less then ϑ then construct a new cluster $C_k = \{d_k\}$; else if there is at least one similarity which is greater than ϑ then d_k is inserted in the existing clusters with a membership degree in the ith cluster:

$$\mu_i(d_k) = \frac{\mathrm{sim}(d_k, v_i)}{\sum_{z=1}^{nbclusters} \mathrm{sim}(d_k, v_z)} \tag{13}$$

3. go back to Step 2.
4. associate to each cluster a set of the q most important key concepts occurring in the documents of the cluster.

Similarity between Document Clusters

We now extend the fuzzy relation sim introduced in Sect. 3.2, which gives the similarity between two document vectors, in order to compute the similarity between two clusters. Let us consider two clusters C_i and C_j and two vector documents x and y. Three methods could then be used for calculating the similarity sim between C_i and C_j:

- *optimal method* in which we use the maximum similarity of pairs. Thus,

$$\text{sim}(C_i, C_j) = \max_{x \in C_i, y \in C_j} \text{sim}(x, y). \quad (14)$$

After merging c_i and c_j, the similarity of the resulting cluster to any other cluster c_k, can be computed by:

$$\text{sim}(C_i \cup C_j, C_k) = \max(\text{sim}(C_i, C_k), \text{sim}(C_j, C_k)); \quad (15)$$

- *pessimistic method* in which we use the minimum similarity of pairs. Thus,

$$\text{sim}(C_i, C_j) = \min_{x \in C_i, y \in C_j} \text{sim}(x, y). \quad (16)$$

After merging C_i and C_j, the similarity of the resulting cluster to any other cluster C_k, can be computed by:

$$\text{sim}(C_i \cup C_j, C_k) = \min(\text{sim}(C_i, C_k), \text{sim}(C_j, C_k)); \quad (17)$$

- *cluster center method* in which we use the similarity between the two cluster centers. Thus, if v_i and v_j are the cluster centers of clusters C_i and C_j respectively,

$$\text{sim}(C_i, C_j) = \text{sim}(v_i, v_j). \quad (18)$$

3.3 Constructing the User Model

The principle of our approach to UMA is the following. When a user logs into the system, there are two possibilities: the user is either new or known. In the former case, the user will be asked to enter her/his personal data and a description of his/her interests (optional). In both cases, an evolutionary algorithm [1], described in Sect. 4, is automatically set to run in the background, its objective being to optimize a model for the current user, with minimal user input.

Information gathered directly from the users at the time of their registration is used to generate, for each user, an initial population of user models. From then on, it is the evolutionary algorithm which is responsible for tracking their interests as they interact with the system. Every action taken by a user in response to any proposition made by the system on the basis of a user model is interpreted as a rating of that user model.

Initializing User Models

The first step, applies to the case of a new user. Our approach consists then of using the set of cluster interests generated as described in Sect. 3.2 to initialize the list of cluster interests in the user profile. Summarizing, a user whose profile does not contain any group of interests (a new user), is asked to describe, with key concepts, document title, phrases or others, his/her center of interest. In the case where this information does not consist of a list of concepts, we consider the list of the most important concepts within the information. These concepts are then compared with the descriptions of all the clusters generated as described in Sect. 3.2. The profile of the user is then set up by adding to the list of its interest groups the instances of clusters with features similar "enough" to those requested by the user. To each of these groups of interests it is associated a degree corresponding to the importance the cluster has to the user. Thus, if $\mathcal{U}$ is the set of key concepts entered by the user and $\mathcal{C}$ the set of key concepts associated to a cluster, the importance $\mathcal{I}$ that this cluster has for the user is given by:

$$\mathcal{I}_{\mathcal{C}} = \frac{|\mathcal{U} \cap \mathcal{C}|}{|\mathcal{U} \cap \mathcal{C}| + |\mathcal{U} \setminus \mathcal{C}| + |\mathcal{C} \setminus \mathcal{U}|} \tag{19}$$

The list of concepts associated to this cluster in the user profile contains only the concepts in $|\mathcal{U} \cap \mathcal{C}|$.

The second step applies to the case when the user is not new. Information about user interests is obtained automatically using a genetic algorithm which optimizes a model whose objective is to propose to the user clusters of documents which are interesting for her. The optimization procedure uses some heuristics described below and a principle of collaborative filtering which consists on updating the user profile by considering also the profiles of similar users.

Summarizing, a user profile contains:

- the personal data of the user,
- a list of fuzzy clusters which interest the user and theirs respective documents,
- the degree with which each cluster interests the user (its importance for the user).

Using Heuristics to update User Models

The user model is automatically updated during the entire acquisition period. Once the initial user profile has been generated, it must be updated. A user profile evolves by means of a genetic algorithm, whose details are given in Sect. 4, with the goal of representing as faithfully as possible the user interests. Below, we describe the methods used for updating the user profile.

The user model is updated during the entire acquisition period because we consider that the user preferences may evolve over time. The typical cases of user interest changes that need to be tracked happen when:

1. the initial clusters are changed, for example, new documents are inserted into some clusters,
2. feedback from the user is available which brings more information about his/her preferences or about changes thereof,

3. there is a user with a similar profile, but richer in terms of information content.

The first point applies to the case in which a new document is added to the repository. The document is then inserted into the cluster to which it is more similar. If the new formed cluster becomes sufficiently similar to an existing cluster in the user profile, it is automatically merged with it.

The feedback of the user allows the system to consider the changes of the user preferences during the acquisition process. Typically it occurs when a proposition based on the model is presented to the user by the system and the user does not agree with all the information reported in it. The feedback is then the result of the interaction between the system and the user.

The third point applies to the case in which there is another user in the system whose profile is sufficiently similar to the profile of the current user. In this case, all the information which is in the similar profile, but not in the profile of the current user, is also inserted into his/her profile. We can extend the similarity measure sim in order to compute the similarity between two users u_1 and u_2 as follows:

$$\mathrm{sim}(u_1, u_2) = \frac{\sum_i \sum_j \min(\mathrm{sim}(i,j), \mathrm{eq}(I(i), I(j)))}{\sum_i \sum_j \mathrm{sim}(i,j)}, \tag{20}$$

where i is the index of clusters in the profile of the user u_1, j is the index of clusters in the profile of the user u_2 and $eq(I(i), I(j))$ returns the equality degree between the importance degree of clusters i and j. Function eq is defined as follows:

$$\mathrm{eq}(x, y) = 1 - |x - y|. \tag{21}$$

4 A Genetic Algorithm for Evolving User Models

Finding, at any point in time, a user model, i.e., the model which allows the system to make the best proposition to the relevant user, is clearly an optimization problem, which presents two main challenges:

1. as user interests may change in time, the problem is dynamic, i.e., its optimum can change and must be tracked;
2. within the scope of our work, a model of a given user can be regarded as a mapping which associates a degree of interest to each document: this general definition would make the search space really huge (the unit hypercube of dimensions equal to the number of documents in the repository), and a more specific definition has to be sought for.

Evolutionary computation [1] provides an ideal technique for approaching this difficult problem, given its ability to effectively search huge solution spaces and track moving optima in dynamical optimization problems.

A genetic algorithm has been implemented for evolving user models within the framework of this work. For each user, the algorithm maintains a distinct population of n models. The population is initialized when the user logs into the system for the first time, and updated each time the user interacts with the system, giving implicit feedback as to the quality of the relevant model. After an initial transition phase, during which the evolutionary algorithm "learns" user interests by trial and error, at steady state, the evolutionary algorithm will maintain a population made of models with a high fitness, and will effectively track any changes in the user interests by having the population evolve accordingly.

4.1 Representation

A user model is represented as a vector of membership degrees in $[0, 1]$, which describe the model's guess at the extent to which the user is interested in each document cluster defined in the system. This choice dramatically reduces the number of degrees of freedom by exploiting the fact that a rational user must have a similar degree of interest for documents having similar contents. Of course, the dimension of such a vector may vary as new clusters are created or a number of existing clusters are aggregated into one bigger cluster; therefore, the evolutionary algorithm will have to take these possibilities into account.

4.2 Fitness Calculation

The evolutionary algorithm is on-line: each time the user interacts with the system, one of the models in the current population is randomly chosen to make its proposition to the user. If the user is satisfied (resp. unsatisfied), the fitness of the chosen model increases (resp. decreases), as well as the fitness of all other models with the same proposition. The fitness of the other models is not changed. The on-line property of our algorithm poses some interesting problems, in that a precise calculation of fitness is not available for every individual at any moment. We invite the reader to see [6] for more details.

Due to the way it is incrementally estimated, the fitness of a model is uncertain. We cope with this uncertainty by representing the (fuzzy) fitness as a possibility distribution over the interval $[0, 1]$, where 0 represents the worst possible rating for a model and 1 represents the rating of a "perfect" model, i.e., a model which always makes the right proposition to the user. The possibility distribution is represented by means of a trapezoid (a, b, c, d), where (a, d) is the support interval and (b, c) is the core interval. When a model is created at random (e.g., when the population is initialized), its fitness is completely uncertain, and thus represented by the possibility distribution assigning a possibility of 1 to all fitness values, namely $(0, 0, 1, 1)$.

Each time a model is tried and the system receives feedback from the user, the possibility distribution (a, b, c, d) representing its fitness is updated to (a', b', c', d') according to rating $r \in [0, 1]$ ($r = 0$ means the user feedback was completely negative, $r = 1$ completely positive) and confidence $w \in [0, 1]$ of the rating, as follows:

$$a' = (a + wr)/(1 + w); \quad (22)$$
$$b' = (b + wr)/(1 + w); \quad (23)$$
$$c' = (c + wr)/(1 + w); \quad (24)$$
$$d' = (d + wr)/(1 + w). \quad (25)$$

4.3 Genetic Operators

The algorithm is elitist, i.e., it always preserves a copy of the best individual unchanged.

The *mutation* operator consists in applying binomially distributed perturbations to each gene independently and with the same mutation probability p_{mut}.

When a model undergoes mutation, the possibility distribution (a, b, c, d) representing its fitness is updated to (a', b', c', d') according to the strength s of the mutation ($s = 0.05n/(n + 1)$, where n is an integer which measures the total perturbations caused by mutation) as follows:

$$a' = \max\{0, b' - (b - a)(1 + s)\}; \quad (26)$$
$$b' = b(1 - s); \quad (27)$$
$$c' = c(1 + s); \quad (28)$$
$$d' = \min\{1, c' + (d - c)(1 + s)\}. \quad (29)$$

The *recombination* operator is uniform crossover: each child has a 0.5 probability of inheriting each gene from either parent. We tried several probability crossover rates and 0.5 is the probability which assures a non homogeneous population after few generations.

When two models are recombined with crossover, the possibility distribution (a, b, c, d) of their offspring is obtained from the possibility distributions (a_1, b_1, c_1, d_1) and (a_2, b_2, c_2, d_1) of the two parents as follows:

$$a = b - (b_1 + b_2 - a_1 - a_2)/2; \quad (30)$$
$$b = \min\{0.95b_1, 0.95b_2\}; \quad (31)$$
$$c = \max\{1.05c_1, 1.05c_2\}; \quad (32)$$
$$d = c + (d_1 + d_2 - c_1 - c_2)/2. \quad (33)$$

Random tournament *selection* with a tournament size of 3 is responsible for selecting the pool of parents for the next generation. This tournament size value has been chosen after many experiments as it is the one which provides an acceptable selective pressure while avoiding premature convergence of the population.

The random tournament consists in extracting a random deviate from a probability distribution obtained from normalizing the possibility distribution of the fitness for each individual in the tournament; the winner is the individual for which the greatest deviate has been extracted.

4.4 Validation of the Approach

The approach described above has been implemented and experimentally validated as follows. A simulation was set up whereby user interests are randomly created and used to simulate user interaction with the system. Precisely, initially a reference model is created. While evolving, the algorithm makes propositions, which are rated according to the reference model. This allowed us to effectively verify that the evolutionary algorithm was capable of learning user interests from user feedback. Figure 1 shows a graph of the distance between the reference model and those used by the system to make its propositions during one of the tests.

In a subsequent phase, a drift effect was applied to the simulated user interests, and the ability of the algorithm to effectively track changing user interests was experimentally proved.

5 Conclusions

The evolutionary algorithm described above will not be always running. It will run only when a request is made by the user. In this case, once a feedback from the user is received, evolution will advance by one generation and then pause, waiting for the next user interaction.

This work is a significant step ahead toward the definition of a UMA approach suited for an ontology-based knowledge management framework, like the one developed in the IKF Project. This approach is novel, although inspired by several works in the literature.

We have first established the representation for the data of the problem. We have redefined the notion of document vector to consider concept, rather than keyword, occurrences in a document. Thus, a document is represented by a vector containing the importance of all the leaf concept obtained when considering the ontology only to a given depth. We have also defined a similarity function for two documents to use a fuzzy clustering algorithm.

A user model is essentially made up of personal data, a set of document clusters grouped according to their similarity and, for each of these document clusters, an associated degree corresponding to the interest of the user in that document cluster. The user models evolve in time tracking changes of user preferences. An evolutionary algorithm has been devised to obtain this behavior. The goal of that algorithm is to find, at each stage in evolution, the best models, i.e., the ones capable of making the best propositions to the user.

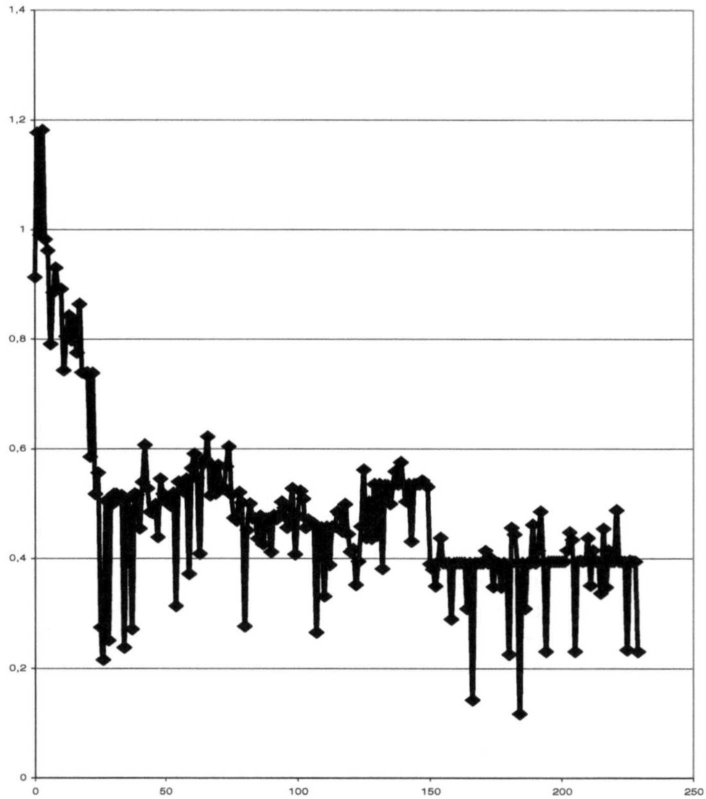

Fig. 1. Distance between the model to learn and the models obtained during the evolutionary process.

References

1. T. Bäck, D. B. Fogel, and T. Michalewicz, editors. *Evolutionary Computation (in two volumes)*. Institute of Physics, Bristol, UK, 2000.
2. Joseph E. Beck, P. Jia, J. Sison, and Jack Mostow. Predicting student help-request behavior in an intelligent tutor for reading. In *Proceedings of the 9th International Conference on User Modeling*, pages 303–312, Johnstown, PA, USA, June 2003.
3. J. Bezdek. *Pattern Recognition with Fuzzy Objective Function Algorithms*. Plenum, New York, 1981.

4. Daniel Billsus and Michael J. Pazzani. Learning collaborative information filters. In *Proc. 15th International Conf. on Machine Learning*, pages 46–54. Morgan Kaufmann, San Francisco, CA, 1998.
5. B. Crabtree and S. Soltysiak. Identifying and tracking changing interests. *International Journal on Digital Libraries*, 2(1):38–53, 1998.
6. E. Damiani, A. Tettamanzi, and V. Liberali. On-line evolution of FPGA-based circuits: A case study on hash functions. In A. Stoica, D. Keymeulen, and J. Lohn, editors, *Proceedings of the First NASA/DoD Workshop on Evolvable Hardware*, pages 26–33, Pasadena, CA, July 19–21 1999. IEEE Computer Society.
7. N.J. Davies, R. Weeks, and M.C. Revett. Information agents for the world wide web. *BT Technology journal*, 14(4):105–114, 1996.
8. IKF Eureka Project E!2235. URL: "`http://www.ikfproject.com/Project.htm`".
9. Christiane Fellbaum (ed.). *WordNet. An Electronic Lexical Database.* The MIT Press, Cambridge, MA, 1998.
10. D. Fisk. An application of social filtering to movie recommendation. In *Software Agents and Soft Computing*, pages 116–131, 1997.
11. C. Goldman, A. Langer, and J. Rosenschein. Musag: an agent that learns what you mean. *Journal of Applied Artificial Intelligence*, 11(5):413–435, 1997.
12. Marko Grobelnik and Dunja Mladenić. Efficient text categorization. In *Text Mining Workshop of the 10th European Conference on Machine Learning ECML98*, Chemnitz, Germany, 21–24 April 1998.
13. Nicola Guarino. Formal ontology and information systems. In Nicola Guarino, editor, *Formal Ontology and Information Systems: Proceedings of the first international conference (FOIS'98)*, Amsterdam, 1998. IOS Press.
14. A. Jain and R. Dubes. *Algorithms for Clustering Data.* Prentice Hall, Dordrecht, 1988.
15. A. K. Jain, M. N. Murty, and P. J. Flynn. Data clustering: A review. *ACM Computing Surveys*, 31(3):264–323, 1999.
16. Mieczysław A. Kłopotek. Very large bayesian multinets for text classification. *Future Gener. Comput. Syst.*, 21(7):1068–1082, 2005.
17. Bernardo Magnini and Carlo Strapparava. Improving user modelling with content-based techniques. In *Proc. 8th International Conference on User Modelling*, pages 74–83. Springer-Verlag, 2001.
18. M. Rodriguez and M. Egenhofer. Determining semantic similarity among entity classes from different ontologies. *IEEE Trans. on Knowledge and Data Eng.*, 15(2):442–456, 2003.
19. Ingo Schwab, Wolfgang Pohl, and Ivan Koychev. Learning to recommend from positive evidence. In *Intelligent User Interfaces*, pages 241–247, 2000.
20. Fabrizio Sebastiani. Machine learning in automated text categorization. *ACM Computing Surveys*, 34(1):1–47, 2002.
21. S. Soltysiak and B. Crabtree. Automatic learning of user profiles — towards the personalisation of agent service. *BT Technology Journal*, 16(3):110–117, 1998.
22. John F. Sowa. *Knowledge Representation: Logical, Philosophical, and Computational Foundations.* Brooks/Cole, Pacific Grove, CA, 2000.
23. M. Steinbach, G. Karypis, and V. Kumar. A comparison of document clustering techniques. In *KDD Workshop on Text Mining*, Boston, MA, August 20–23 2000.
24. Frank Wittig. ML4UM for bayesian network user model acquisition. In *Proceedings of the second workshop on machine learning, information retrieval and user modeling*, June 2003.

25. Yiming Yang, Tom Ault, Thomas Pierce, and Charles W. Lattimer. Improving text categorization methods for event tracking. In *SIGIR '00: Proceedings of the 23rd annual international ACM SIGIR conference on Research and development in information retrieval*, pages 65–72, New York, NY, USA, 2000. ACM Press.
26. Y. Zhao and G. Karypis. Evaluation of hierarchical clustering algorithms for document datasets. *ACM Press*, pages 515–524, 2002.
27. Ting Shao Zhu, Russell Greiner, and Gerald Häubl. Learning a model of a web user's interests. In *Proceedings of the 9th International Conference on User Modeling*, pages 65–75, Johnstown, PA, USA, June 2003.

On Some Problems of Decision–Making Under Uncertainty in the Semantic Web

Miguel-Ángel Sicilia

Computer Science Department, University of Alcalá
Ctra. Barcelona km. 33.600 – 28871, Alcalá de Henares, Madrid (Spain)
msicilia@uah.es

Summary. The Semantic Web is a concrete meta-technological framework in which an existing infrastructure of hypermedia elements and distributed services is provided with meta–data describing them in a loosely–coupled way. This entails that software agents are faced to reasoning and acting on a domain in which possible worlds are fundamentally prescribed by the metadata records available. Nonetheless, metadata records can be used to express different forms of information imperfection, and they may in some cases be conflicting or inconsistent when contrasted. This leads to a number of problems that should be addressed in the transition of the view of the Semantic Web as a unified and "crisp" knowledge representation to a more flexible reasoning framework that considers uncertainty and imprecision as a first–class characteristic. This paper provides an account of some important problems that arise in such flexible account of the Semantic Web, in an attempt to outline a context for further research.

Key words: Semantic Web, ontologies, uncertainty

1 Introduction

The *Semantic Web* [3] can be considered as a *meta-technological* framework that opens up a whole new range of solutions for the tasks of delivery and distribution of information through the World-Wide-Web, and that provides the substrate for a new space of virtual common culture [16, 39]. This is due to the fact that it provides a technology that is able to enhance the existing Web technology by adding *sentences* in formal, machine-understandable languages to describe the elements of the Web. That is, annotations or semantic metadata conform a layer of description referring to the actual Web elements (Web pages, services, images, etc.). The current Web architecture is based on the concept of *hypermedia*, recently extended by the *service-oriented* architecture concept, which now covers *Grids* and on-demand computing elements

M.-Á. Sicilia: *On Some Problems of Decision–Making Under Uncertainty in the Semantic Web*, StudFuzz **204**, 231–246 (2006)
www.springerlink.com

also [7], so that these are the fundamental objects described by the sentences of Semantic Web applications or systems.

As a meta-technology based on knowledge engineering, Semantic Web techniques should represent and reason about the current Web ontology. This entails that some form of representation of both hypermedia and services is a pre-requisite for enabling the predication of sentences about Web resources[1]. The facilities provided by recent W3C-endorsed languages as XPath[2] and XPointer[3] provide the standardized language for the description of Web elements of an arbitrary granularity, and the many existing hypermedia models — currently extended in the form of open hypermedia models [9] — allow for the modeling of hypermedia form a highly abstract perspective. Further, initiatives as OWL-S and WSMO deal with the description of services [14]. Consequently, the representation of the material object to be described seems at first glance attainable by current means.

If we turn our attention to the form of the sentences in the Semantic Web, the current technology is (arguably) essentially based on description logics [2], which in turn are essentially a family of logics for which we know their computational properties. If we tackle the problem of the Semantic Web from a reductionist viewpoint, the base technology appears to be ready and the use of shared and publicly available ontologies [6] would resolve the issue of having a common knowledge representation. From this viewpoint, ontology integration [17] complements the "consensus" approach with techniques to deal with ontology alignment. Many Semantic Web applications are based on this presuppositions, and in fact many applications that operate inside organizational boundaries or that are restricted to a subset of controlled Web resources can safely assume them.

Nonetheless, the emergence of the Semantic Web is not as simple as it appears at first glance, and description logics and their related languages as OWL require extra–logical considerations to tackle with issues related to basic problems of epistemology as *trust* [8]. This comes from the nature of the WWW as a decentralized and open system, properties that are considered by many as the key of its success. In other words, if we assume that Semantic Web agents will benefit from sentences not generated in a closed domain but created in a distributed and open fashion, then an architecture for shared uncertainty representation is required. Without it, Semantic Web applications will be restricted to some pre-established knowledge and information sources, thus not leveraging the open and distributed nature of the Web.

Veltman [39] described some limitations of the existing Semantic Web approach, including the the assumption of a *single world–view* on knowledge representation, precluding divergent and changing paradigms, and other

[1] Here the term "resource" is used in a generic sense, including pages, contents, fragments of markup, XML-based descriptions and services, among others.

[2] http://www.w3.org/TR/xpath

[3] http://www.w3.org/TR/xptr/

related to meanings and language. This is an important point, for which microtheories (modules of sentences that require internal consistency but may be inconsistent with other microtheories) as those used in the *Cyc* project [15] could provide a solution. But the issue of imperfection [32] (including uncertainty, imprecision and inconsistency) in the sentences has not been mentioned explicitly in existing critical essays, and this broadens the single world-view problem to a more general framework. Further, information imperfection is a cross–cutting characteristic for many domains [27], so that general techniques are required for the sake of "semantic interoperability".

Probabilistic and fuzzy approaches to extending description logics provide numerical approaches to deal with "uncertainty" in existing work, e.g. [35, 33], but they do it without an explicit examination of the underlying epistemological issues. This causes many open problems for current technology. For example, let's suppose a (semantic) commerce site is acting autonomously to recommend cars to one of its users. Then, the system could have the following knowledge (this is adapted and expanded from the example in [34] and uses a variant of description logics):

$$SportsCar = Car \sqcap \exists speed.very(High)$$
$$High(x) = min(1, 0.004x)$$
$$very(x) = x^2$$
$$fa - car : Car \sqcap (\exists maker.renault) \sqcap (\exists speed.\{243km/h\}) \sqcap (\exists model.\{rt100\})$$
$$wp : ProductDescriptionPage \sqcap (\exists describes.\{rt100\} \geq 0.6)$$
$$user : Users \sqcap (\exists interestInSportCars.30)$$

The following problems related to uncertainty and imperfection arise even from such simple situation:

(a) The fact that a sports car is a car with high velocity seems an uncontroversial, inter–subjective piece of knowledge. However, how was the membership function for the *High* label determined? how can the software determine the "right" numerical function for the term? does the fuzzy concept of a fast car evolve with the course of the years? The same applies to the determination of the mathematical function corresponding with the hedge *very*. Is it an universal, uncontroversial definition?
(b) An inference using fuzzy description logics can derive the assertion $fa - car : SportsCar \geq 0.94$. Let's suppose this is found by another software that does not know the origin of the assertion. This means that is there uncertainty (with the available evidence) about *fa-car* being a sport car? or does it mean partial membership?
(c) What should the agent do if it finds on the Web a conflicting assertion, let's say $fa - car : (\exists speed.160km/h)$? How can it decide which information is more credible?

(d) The interest of the individual in sport cars is uncertain since interest can be measured only through indicators. How could we compare indicator–based measures without common scales?
(e) With the information available, should the agent show the page to the user? In this case, the results of the action could not entail negative results (despite annoying the user), but what if the decision would be that of initiating an order, which has a more concrete economic effect?
(f) For the assertion that describes the Web page *wp*, what is the shared vocabulary that describes Web pages and other resources? how could we interpret the degree showed in the assertion (uncertainty, imprecision, strength of association)?

Some questions as the notions of trust deriving from problem (c) are currently focus of ongoing research. For other questions techniques exist to deal with them, but there is not a systematic effort or interest in them. This is the case of (a), for which reliable inter–subjective representations can be obtained from membership elicitation techniques [4], but few studies actually exist.

The questions above will be dealt with in the rest of the chapter.

As the discussion above evidences, there is a need to reflect on the kind of uncertainty agents or software modules face in their routine decision making on the Web. Symbolic description logics are being extended with probability or fuzzy–based numerical approaches, but the logical machinery is not enough if the epistemological and interpretive issues are not carefully examined and put in a shared form in some way. This chapter attempts to describe some of the general issues of uncertainty (understood as a broad term) in the context of the Semantic Web, in an attempt to foster research on shared epistemological frameworks for future semantic technology. The general principles proposed accept many different technological solutions, so that further work should deal with a process of rational inquiry on the best of them.

The rest of this chapter is structured as follows. Section 2 provides an overview of the object of Semantic Web descriptions. Then, Sect. 3 describes a number of concrete paradigmatic uncertainty-related problems that arise in the context of the Semantic Web. Section 4 outlines the main requirements for a future technological framework that integrates the problems described. Finally, conclusions are provided in Sect. 5.

2 An Ontological Account of the Web as an Hypermedia System

This section briefly sketches the main elements of the ontology of the Web as the object of Semantic Web services, to provide an overview of the main classes of digital entities that can be described through metadata. A more detailed account can be found in other papers as [30] and [1].

2.1 Core Abstract Components of the Web as a Hypermedia Structure

Hypermedia structure can be succinctly expressed in terms of a small number of concepts, namely, nodes, contents, anchors and links. Here an account of them (with some links to *OpenCyc*[4] definitions to improve the meaning of the definitions) are provided. These are discussed in more depth in [30].

Nodes in hypermedia considered as "containers of information" of different media types. This definition is closely connected with the *OpenCyc* notion of `InformationBearingThing`, and even with its specialization `ComputerFileCopy` if a more restrictive view on the "storage" layer [10] of hypermedia elements is considered. The notion of "*content*" is usually also defined as "a piece of information". The subtle difference lies in that nodes can be arbitrarily composed while contents are considered as atomic pieces of information with a specific type (audio, video, etc). In addition, the consideration of what is a node is somewhat conventional, since it is a unit of navigation of the hypermedia space, i.e. nodes are rendering units in the user agent (browser). Nodes can be arbitrary aggregations of other contents, so that their identity is provided by their configuration.

Links are first–class citizens in hypermedia models, contrary to the realization of links in the Web, where they are embedded in nodes (i.e. pages) as concrete markup elements. Considering links as independent entities allows the definition of multi-directional, complex relational structures, as can be defined in *XLink*. In addition, links can also be used as assets in themselves, since they may embed type information [36]. The description of full–fledged links can be derived from the XLink specification in terms of *locators* and *arcs* that conform graphs with edges labeled by types. Therefore, links depend on the locators they connect, i.e. through the notion of *Anchor*. It should be noted that a concrete user–interface related interpretation of both multi–dimensional links and link types is not currently agreed. This fact could also be represented in the hypermedia ontology as annotations to the link–describing elements themselves.

In addition to the described elements, users and groups have for long time be considered as first-class elements of the hypermedia space, and as such they. The *OpenCyc* term `IntelligentAgent` provides a general account of users (artificial and human) capable of knowing and also for collectives as `Organizations`. Further, links can also be considered as information bearing things just as contents and nodes. The difference lies in their relational nature.

2.2 Metadata About Services

The WSMO framework provides a convenient language for the description of services in general, and of Web Services in particular. Without going into

[4] http://www.opencyc.org

details of the details of this kind of technology, the main conceptual elements of such descriptions are the following [1]:

- *Capability*: Description of a Web Service by means of its functionality, based on pre-conditions, post-conditions, assumptions and effects.
- *Goal*: Specifies the objectives that a client may have when he consults a web service based on post-conditions and effects. A goal can be made of constituent sub-goals.
- *Interface*: Specification of how the functionality of the service can be achieved, by means of fulfilling its capability. It takes a twofold approach based on choreography, or how to communicate with the web service in order to consume its functionality, and orchestration which defines how the overall functionality is achieved by the cooperation of more elementary service providers.
- *Non-functional properties*: Properties related to quality aspects of Web Service (QoS).

Goals and capabilities are the two main elements in which information imperfection is a key issue. Arbitrary sentences may be used for their description, since they are not restricted to an input–output data paradigm, but they are sentences on arbitrary domain ontologies — which of course could include descriptions of Web resources as nodes or links as those described in the above sub–section.

From an ontological viewpoint, services as programs running can be assimilated to *OpenCyc* `ComputerProcessRunning` instances. However, further definitions are required for a comprehensive modeling of Web Services as described by current technology.

3 Some Paradigmatical Problems of Reasoning with Uncertainty in the Semantic Web

This section describes a number of paradigmatical problems that are specific to the characteristics of the Semantic Web as an extension that preserves the open principles of the Web.

3.1 Some of the Few Objective Hypotheses Possible

Ontology as a formal engineering artifact is essentially subjective (or better, inter–subjective) in terms of its epistemological status, according to the sense of the term provided by Popper [19]. That is, the assertions and conceptual structures provided by ontologies and the instances of their terms are not subject to "falsification", they are a form of tautological knowledge that relies in authority and "consensus" as its source of validity. This gets even worse if we admit that assertions on the Semantic Web could be created by any individual. Thus, the determination of the classes of validity of Semantic Web knowledge is critical for a realistic technology.

A class of objective statements can be defined as those for which there are procedures that can be used to test it. This makes that kind of statements subject to falsification, so that agents have a chance to test them. However, the class of this statements in the Semantic Web is rather limited, and they affect mainly to sentences on the availability of resources, the structure and behavior of Web elements or services or some restricted forms of checking that can be tested by relying on some elements of the Web structure. For example, the assertion $p : HTMLPage \sqcap (\exists uri.\{http : //www.uah.es\})$ can be simply tested by requesting the Web page and parsing its contents. Probabilistic assertions can also be checked, as for example, testing that the probability or propensity of users to follow a navigational path in a concrete Web site is greater or equal to 0.8, which are common hypotheses in Web usage mining [13].

Further, axioms like $PagesInDomainA \sqsubseteq HTMLPage$ can be checked through similar means. The common element of such assertions is that they deal with the physical, digital part of the hypermedia structure, so that they do not deal with domain or commonsense knowledge, which are by their definition tautological or subjective, i.e. they are not subject to objective test. Algorithms like Google's *PageRank* [18] operate only on such objective structural elements, even though the rankings produced are of course imprecise and subject to interpretation. Another relevant category of objective measures are that of social network analysis as applied to digital communications or structure [28], even though once again, objectivity only refers directly to the numerical models, and not to their interpretation on social terms except if objective correlation studies have been carried out.

If we move from these kind of objective assertions to others, the consideration of validity changes radically. Let's examine an apparently uncontroversial situation as the following. An agent queries a Web Service provided by the company IBM that provides contact information about its employees, searching for "John Doe". The information gathered from the service tells that John left the company about mid-1998 (let's suppose there was a failure in the accounting records, so that the exact date is not know) which could be encoded as $leftDate : FuzzyTemporalEvent \sqcap (\exists organization.\{ibm\}) \sqcap (\exists person.\{johndoe\}) \sqcap (\exists approximate - time.\{mid - 1998\})$. Even though we will rarely doubt of the validity of such information, its validity is not contrastable without reverting to any form of *authority* or *trust*. Hence, if the agent takes the information as true, it does so since it considers IBM as a credible source[5].

[5] Of course that the possibility of a security break could make the agent doubt that the response came from IBM actually, but this can only be solved by technical security solutions.

In the World Wide Web, the credibility of contents is left mostly to the judgement of the human reader, which has the potential ability to discriminate to some extent and at least in some situations. So, software agents require the provision of trust services or trust–assessing mechanisms of any kind to perform even the more simple tasks. An that trust model is not an accidental or optional feature, but a central element in the Semantic Web architecture that is in many cases overlooked. In fact, many current applications use as a trust model the simple fact that they only rely on knowledge inside a particular server or ontology, but this is not coherent with the vision of the Semantic Web as a global, open knowledge infrastructure. Rather, it is simply the application of AI techniques to restricted domains.

3.2 The Interpretation of Fuzzy Assertions and Axioms

The extension of symbolic logics to numerical frameworks entail that a new interpretation should be given to the numbers appearing with the assertions and axioms. There is not a single framework for the interpretation of the various kinds of information imperfection [32], so that there is a need for a representation of the diverse representations of uncertainty. Given that there exist some general frameworks for several uncertainty models (including probability) [11], it seems reasonable to have some general representation of the different models (i.e. an ontology of mathematical models for uncertainty) so that agents can gain access to the algorithms that should be used for each of them, and are also aware of which are compatible knowledge pieces and which not.

For imperfection in knowledge, there are some previous works on cognitive psychology that could be used as models for it. Notably, the research conducted by Rosch and her colleagues on human categorization [20] is in surface compatible with the main original postulates of fuzzy logics. Categorization is viewed in such standpoints as the process of grouping things based on *prototypes*, i.e. the idea of necessary and sufficient conditions is rarely if ever met in categories of naturally occurring things. Systems of categories are thus not fixed and external to human beings but they are rooted in people's experience. But even though conceptual categories are not identical for every speaker of a language, there is some inter-subjective perception, that could be mediated by culture but that provides some coherence to human categorization [25]. For that models to be *shared* together with ontologies, there is a need to integrate membership function elicitation and similar experimental techniques into the ontology engineering process [38]. This would end with public definitions in the general form (1).

$$\mu_{C,ctx} : D_1, D_2, \ldots D_n \rightarrow [0,1] \tag{1}$$

Expression (1) entails a considerable complexity since the characterization of a concept C depends on a context ctx, and the specification of such context can be a complex formula or decision procedure. Further, the domains D_k used for the determination of membership can be properties of any kind and even values coming from other procedures. The computation procedures can be encoded in the form of a Web Service in the most complex cases for public access. The context here is a placeholder for any cultural or domain–specific use of the concept. Since concepts are deeply connected with our natural languages, several perceptions of similar concepts can be found in different countries, and also some general concepts can be interpreted differently in different situations.

Without that, fuzzy description logics will entail a degree of arbitrariness in the definitions of operators that is not coherent with the aim to model human knowledge "consensually", and not specifically for a given small problem. Even though fuzzy set theory is intuitively based on such cognitive hypotheses of human characterizations, the relationship of the purely mathematical models with the cognitive structures they are supposed to model has received little attention. Furthermore, membership function elicitation studies have shown that the mathematical models for some hedges do not apply to every situation [4]. Even the operators that are applied to the computation of concept intersection and union, and existencial or universal quantification are subject to empirical adjustment or validation for some situations [26]. Thus, a huge effort and new cost–effective methods are required for the representation of vagueness as associated to inter–subjective categories. Another element in which vague prototype category theories can be applied is on the "distance" of generalization-specialization relationships, that in crisp logics are considered as bi-valued. In that case, it is easy to obtain evidence about the fact that some of these relations are considered by humans as closer than others. For example, in a e-commerce application, some sub-classes of products can be more similar to the immediate "super-class" than others. A first tentative model for this kind of resemblance relations is described in [23].

In any case, it seems reasonable that representations of the models are required, and that every derived (i.e. non–primitive) assertion or concept definition that entails a form of numeric imperfection is linked to such explicit representations.

3.3 Fuzziness in Hypermedia Navigation

The dynamics of the Web as a hypermedia structure are centered around the concept of navigation, which has been subject to study as an enhancement or facilitator of some cognitive human processes [21]. This entails that not only concepts in the contents of the pages (nodes) should consider vague models, but also the relationship between nodes and contents that are established by links. A model for such vague linking has been described in [29], but again, shared and publicly available models should be made available in the Semantic Web. In fact, the work of Trigg [36] was seminal in research on link typing, so that there is a form of commonly accepted proto-ontology for the semantics of links (but restricted to formal writing).

Hypermedia links create relationships between two or more addressable items or contents with a concrete semantic or rhetoric intention. If that intention is made explicit, it can be used by the hypermedia system to provide the link with adaptive presentation, hiding or rearranging for specific users [5]. In consequence, ontologies of link types may serve the purpose of codifying common link intentions in a formal way (e.g. as a taxonomy of link types derived from a *Link* concept). But a given node n_i can be connected by link L to a number of destination nodes n_j with varying degrees of strength, as described in [29]. For example, a *DetailLink* may be linked with an strength of 0.8 to a full explanation of a content item, while having a link of strength 0.3 to a brief explanation. This way, the navigation structure itself becomes fuzzy, and imprecise relations can be used to implement diverse adaptations of the UI, e.g. sorting destinations by degree, or filtering only the most relevant ones. In [29], the concept of *fuzzy link* was proposed as a model for such interface elements, expressed in (2), where S and T are the sets of sources and targets (addressable entities), respectively, U is the set of users, and Λ is a set of membership functions.

$$L = (S, T, \Lambda) \;\; and \;\; \mu_{(L,p)} : (S, T, U) \rightarrow [0, 1] \;\; with \;\; \mu_{(L,p)} \in \Lambda \qquad (2)$$

Here imprecision is essentially attached to each concrete link, but the problem of interpretation of the numerical assessments appears just in another form. Again, the concrete computational semantics of such numbers should be made explicit and available to software agents.

3.4 The Accumulation of Evidence and Trust Mechanisms

Another important aspect of Semantic Web technology is the value assigned to the accumulation of evidence. Some uncertainty frameworks deal explicitly with bodies of evidence [11], but in the Semantic Web this accumulation refers to a situation of social trust, for which paradigmatic examples are those of the Wikipedia or other "open" efforts.

In many cases, the problem of accumulation of evidence is related to the concept of "consensus". Ontologies themselves are considered by many as "consensual" representations, and the idea that open "consensus" has some epistemological validity is reflected in the many "rating-based" assessment applications in the Web, including collaborative filtering [12]. Nonetheless, consensus may be considered as a potential indicator, but not as a definitive validity criterion. For example, if many individuals provide values for z in an assertion about an educational Web resource as $p : LearningObject \sqcap (\exists expostion.\{active\} \geq z)$, there are several criteria that can be applied to determine the numerical, inter-subjective value of the assertion. But such criterion are of a different degree of objective credibility. In many cases, known statistical techniques of rater agreement [37] can be used for that purpose. For example, when using Likert-type scales (i.e. those that assume equal distance between possible ratings), factor analysis can be used to deliver an estimation of divergences with respect to one or several "latent traits". However, these procedures require some form of interpretation, thus requiring rational inquiry processes [24] that are seldom found in current approaches to intelligent systems operating on ratings.

The result of the social consideration of consensus on knowledge is a subtle form of trust assessing mechanism, in which the source evaluated is not a concrete entity, but a general hypothesis on the validity of consensual approaches. Consensus evaluation could also be applied to non–numerical knowledge, but that poses an even more challenging restriction, since there are not clear measures of agreement for such elements.

3.5 Decision Making and the (Human) Value of Actions on the Web

A basic assumption of the Semantic Web is that it will increase the degree of *delegation* of tasks to agents. If we adhere to the definition of the Merriam-Webster dictionary for the term, this will mean "empowering the agent to act for the user it represents". A typical example of this empowering may be that of enabling an agent property of organization A to engage in business processes with services available in other organizations. In practical terms, this is in general a form of decision–making based on risk. The underlying element of risk assessment in most cases is that of evaluating subjective probabilities of taking the wrong action with respect to some *value* [31]. Value is often measured in economic impact on a project or organization, but other elements could also be considered as impact on the credibility of an application. In any case, the notion of value is contingent to humans or organizations, so that an agent operating on them should be equipped with a computational representation of them. In turn, and given the uncertain nature of risk, such representations must be imprecise and numerical. Further, the Semantic Web environment poses the challenge of assessing risk in terms of a diversity of possible information imperfection sources. In consequence, all the elements commented above should be taken into consideration in the final act of automated decision making.

The discussion above evidences that a realistic consideration of uncertainty and imprecision in a open Semantic Web environment requires a complex structure of representations, far beyond the simple use of standard domain ontologies in OWL. In what follows, an attempt to provide an architecture and some principles for such framework are provided.

4 Towards Shared Uncertainty Representations

The above discussion provides a picture of the variety of problems that are required to deal with in the context of a Web with an open infrastructure, in which a central authority for metadata does not exist. The archetypical problems described illustrate the different aspects that an agent or software module would have to tackle with in its routine processing. Figure 1 provides an architecture that integrates all the aspects considered.

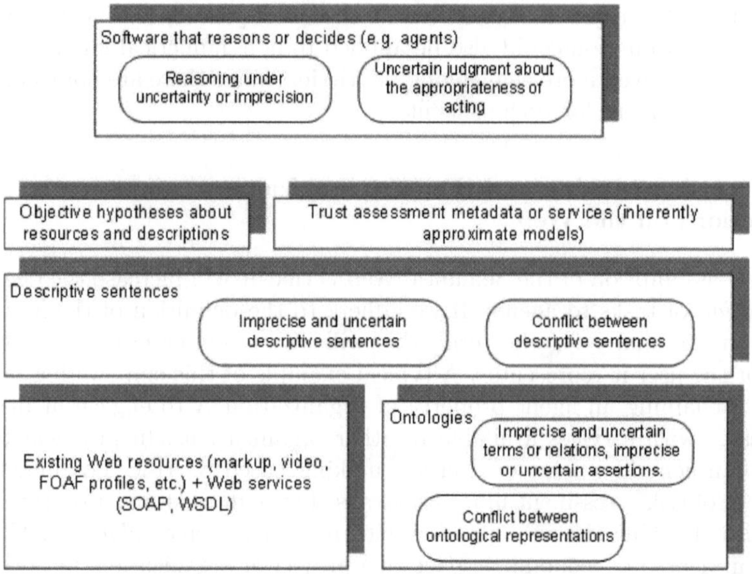

Fig. 1. Overall view of uncertainty on the SW.

The current ontology of hypermedia and Web services as described in the second section of this chapter is represented at the bottom of Figure 1. At the same level, ontology libraries are expected to be publicly available, but including numerical uncertainty and imprecision representations, with explicit computational models for the different mathematical frameworks. At this level, inconsistency [32] in the represented knowledge should be dealt with, possibly including consensus–based imprecise assessments.

A second level in Figure 1 represents the descriptions that are the actual core of the Semantic Web. These descriptions will predicate about hypermedia elements or Services, using the ontologies for that. Here a second kind of inconsistency may also appear.

The third level in Figure 1 is critical in that it adds a consideration of the validity of the information in the lower layers, in a form of trust assessment. Such information could be used by agents or software modules to discard or give less credit to some descriptive sentences or to some ontology definitions or assertions. That combined system of believability and evidence is the raw material for decision making. The last component is the consideration of value as the final criteria to act.

Departing from the general vision depicted so far, a number of principles can be stated for research on a "Uncertain Semantic Web". Some of them are discussed in the remainder of this section.

Principle 1 *Numerical fuzzy assertions, concept or predicate definitions must be accompanied by the semantics of their imperfection, that is, by the kind and/or source of imperfection they represent.*

Principle 1 entails that some kind of ontology of information or knowledge imperfection types is required as a basic building block. With such representations, agents may be able to apply specific paradigms of reasoning under uncertainty that are consistent with the origin or the purpose of the sentences they are dealing with. This will alleviate problem (b) as stated above.

Principle 2 *Just as common ontology engineering attempts to provide consensual, inter-subjective knowledge representation, the vagueness of concepts, modifiers and predicates should be measured and elicited by rigorous inter–subjective procedures.*

Principle 2 complements principle 1 with an emphasis on empirical methods or rigorous procedures for the determination of vague concepts, thus avoiding "intuitionism" of individuals or divergence of representations. Since vagueness is pervasive in many Web applications [22], this would be critical for shared understanding under imprecision. This focuses on solving problem (a) as stated in the introduction.

Principle 3 *Trust and reputation mechanisms as those that are proposed in current research should be extended to deal with imprecise or uncertain degrees of trustworthiness or credibility, which are consequence of the imperfection or partial truth of claims.*

Principle 3 extends the consideration of imprecision and uncertainty to the mechanisms of trust and reputation, thus providing a flexible mechanism to deal with problems of inconsistency as the one raised in problem (c).

Principle 4 *The relationship of actions taken from decision under uncertainty should be based on some model of their value to the organizations or people that "own" the agents or software modules. This would in fact be a model of risk in decision making, and it is uncertain by its own nature.*

Assumption 4 tackles with problem (e), but it is determinant of the final usefulness of the technology, since it defines the degree of actual "delegation" that software would fully automate.

These set of assumptions lead to a concept of "shared uncertainty (or imprecision) representation" that represents a shift from the current crisp–logics dominant paradigm in the Semantic Web. However, the feasibility of such representation paradigm (as compared with the current one) is still a subject for further inquiry, since it requires the combination and organization of a large amount of mathematical models combined with the collaborative development of empirical studies for the determination of inter–subjective "fuzzy ontology" representations. This represents an avenue for research and development still largely unexplored.

5 Conclusions

The Semantic Web can be considered a meta–technology focused on the description of existing Web resources for the concrete purpose of enabling the delegation of tasks to software agents. Current Semantic Web technology relies on Web–ready ontology description languages as OWL, which in turn builds on research on description logics. Information imperfection is a cross-cutting aspect for any domain of application, and the open and distributed nature of the Web additionally raises a number of questions on the handling of uncertainty and imprecision. These questions should be addressed for realistic applications to operate in a open environment, not restricted to closed computing environments.

Trust, shared representations of imprecision and an explicit representation of types of information imperfection can be used as the foundation for consistent agent behavior. Such consistency should be complemented by some representation of the value of actions for humans or organizations, which would determine the propensity to ask in a form of decision making under risk. Such layered architecture has been sketched in this paper as a complement to other technical proposals yet commonly accepted. Such layered view can be used for roadmapping in approaches to a "fuzzy semantic Web" that exploits the many techniques coming from Soft Computing research into a coherent and shared framework. Without such shared and comprehensive approach, only scattered and isolated applications could be devised, thus loosing the opportunities provided by the open and public infrastructure of the Web.

References

1. Arroyo, S. and López–Cobo, J.M. (2005) Describing Web Services with Semantic Metadata. *International Journal of Metadata, Semantics and Ontologies* 1(1).
2. Baader, F., Calvanese, D., McGuinness, D., Nardi, D., Patel-Schneider, P. (eds.). (2003). *The Description Logic Handbook. Theory, Implementation and Applications*, Cambridge.
3. Berners-Lee, T., Hendler, J., Lassila, O. (2001). The Semantic Web. *Scientific American*, 284(5): 34–43.
4. Bilgiç, T. and Türksen, T.: Measurement of Membership Functions: Theoretical and Empirical Work. In: D. Dubois and H. Prade (eds.) *Handbook of Fuzzy Sets and Systems* Vol. 1, Chapter 3, Fundamentals of Fuzzy Sets, Kluwer (1999):195–232

5. Brusilovsky, P. (2001) Adaptive hypermedia. *User Modeling and User Adapted Interaction* **11**(1/2): 87–110
6. Ding, Y. and Fensel, D. (2001). Ontology Library Systems: The key for successful Ontology Reuse. *Proceedings of the first Semantic web working symposium* (SWWS1), Stanford, USA
7. Foster, I. and Tuecke, S. (2005). Describing the elephant: the different faces of IT as a service, *ACM queue* 3(6): 26–43
8. Golbeck, J. and Parsia, B. (2005) Trust network-based filtering of aggregated claims. *International Journal of Metadata, Semantics and Ontologies* 1(1)
9. Grønbæk, K. and Wiil, U. (1997) Towards a Common Reference Architecture for Open Hypermedia. *Journal of Digital Information* 1(2)
10. Halasz, F. and Schwartz, M. (1994). The Dexter hypertext reference model. *Communications of the ACM*, 37(2):30–39
11. Klir, G., Wierman, M. (1998) *Uncertainty-based information*. Springer-Verlag, New-York.
12. Konstan, J., Miller, B., Maltz, D., Herlocker, J., Gordon, L., and Riedl, J. (1997). GroupLens: Applying Collaborative Filtering to Usenet News. *Communications of the ACM* 40(3): 77–87
13. Kosala, R. and Blockeel, H. (2000). Web Mining Research: A Survey. *SIGKDD Explorations*: Newsletter of the Special Interest Group (SIG) on Knowledge Discovery and Data Mining, ACM, ACM Press: 1–15
14. Lara, R. and Polleres, A. et al. (2005). A Conceptual Comparison between WSMO and OWL-S. *WSMO Working Draft*, January 6, 2005, available at http://www.wsmo.org/2004/d4/d4.1/v0.1/
15. Lenat, D. B. (1995). Cyc: A Large-Scale Investment in Knowledge Infrastructure. *Communications of the ACM* 38(11): 33–38 .
16. Mitcham, C. (1995). Notes Toward A Philosophy Of Meta-Technology. *Techné* 1(1-2)
17. Noy, N. (2004). Semantic integration: a survey of ontology-based approaches. *SIGMOD Record*, 33(4)
18. Page, L., Brin, S., Motwani, R. and Winograd, T. (1998) The PageRank citation ranking: Bringing order to the web. *Technical report, Stanford Digital Library Technologies Project*. Paper SIDL-WP-1999-0120 (version of 11/11/1999).
19. Popper, K. (1985) Realism and the Aim of Science (Postscript to the Logic of Scientific Discovery), Routledge.
20. Rosch, E. (1978) Principles of Categorization. In: Rosch, E., Lloyd, B. (eds.), *Cognition and Categorization*, Lawrence Erlbaum, Hillsdale: 27–48
21. Rouet, J. (editor). (1996) *Hypertext and Cognition*. Lawrence Erlbaum Publishers, New Jersey.
22. Sicilia, M.A., García, E., Díaz, P., Aedo, I. (2002) Fuzziness in adaptive hypermedia models. In: *Proceedings of the North American Fuzzy Information Processing Society Conference*: 268–273.
23. Sicilia, M.A., García, E., Díaz, P., Aedo, I. y Díaz, P. (2003) Representation of Concept Specialization Distance through Resemblance Relations. *Advances in Soft Computing – Engineering, Design and Manufacturing* XII, Springer: 173–182
24. Sicilia, M.A. (2003) Observing Web Users: Conjecturing and Refutation on Partial Evidence. In: *Proceedings of the North American Fuzzy Information Processing Society Conference*

25. García, E., Sicilia, M.A., Gutierrez, J.A. (2003) On the Vague Modelling of Web Page Characteristics Regarding Usability. *Web Intelligence LNAI* 2663, Springer: 199-207
26. Sicilia, M.A. (2003) The Role of Vague Categories in Semantic and Adaptive Web Interfaces. *Human Computer Interfaces for Semantic Web Applications LNCS* 2889, Springer: 210–220
27. Sicilia, M.A., García, E. (2004). On imperfection in information as an "early" crosscutting concern and its mapping to aspect-oriented design. In P*roceedings of Early Aspects 2004: Aspect-Oriented Requirements Engineering and Architecture Design Workshop*
28. Sicilia, M.A., García, E. (2004) Fuzzy Group Models for Adaptation in Cooperative Information Retrieval Contexts. *Lecture Notes in Computer Science* 2932 Springer: 324-334.
29. Sicilia, M.A., Garcia, E., Aedo, I. and Diaz, P. (2004). Using links to describe imprecise relationships in educational contents, *International Journal of Continuing Engineering Education and Lifelong Learning* 14(3): 260–275.
30. Sicilia, M.A., García, E. (2005). Information System Aspects and the Ontology of Hypermedia Systems. In Rajiv Kishore, Ram Ramesh, Raj Sharman (eds.) *Ontologies in the Context of Information Systems* (Springer, to appear).
31. Sicilia, M.A., Lytras, M. (2005) The semantic learning organization. *The learning organization* 12(5): 402-410
32. Smets, P. (1997). *Uncertainty Management in Information Systems: From Needs to Solutions*, Kluwer, chapter Imperfect information: Imprecision- Uncertainty: 225–254.
33. Straccia, U. (2001) Reasoning within fuzzy description logics. *J. Artificial Intelligence Research*, **14**: 137–166.
34. Straccia, U. (2005). Towards a Fuzzy Description Logic for the Semantic Web (Preliminary Report). *Lecture Notes in Computer Science* 3532, Springer: 167–181
35. Tresp, C.B., Molitor, R. (1998) A Description Logic for Vague Knowledge. In: *Proceedings of the 13th biennial European Conference on Artificial Intelligence* (ECAI'98), J. Wiley and Sons, Brighton, UK 361–365
36. Trigg, R. H. and Weiser, M. (1986). TEXTNET: A Network-Based Approach to Text Handling. *ACM Transactions on Office Information Systems*, 4(1): 1–23
37. Uebersax, J. S. (1992). A review of modeling approaches for the analysis of observer agreement. *Investigative Radiology*, 27 :738–743.
38. Uschold, M. and Gruninger, M. (1996) Ontologies: principles, methods and applications. *Knowledge Engineering Review*, 11:2: 93–136
39. Kim H. Veltman (2002) Towards a Semantic Web for Culture. *Journal of Digital Information* 4(4)

Automatic Thematic Categorization of Multimedia Documents using Ontological Information and Fuzzy Algebra*

Manolis Wallace, Phivos Mylonas, Giorgos Akrivas, Yannis Avrithis, and Stefanos Kollias

Image, Video and Multimedia Laboratory, School of Electrical and Computer Engineering, National Technical University of Athens, Greece
wallace@image.ntua.gr

1 Introduction

In the framework of digital archiving of multimedia content different scientific fields such as database management systems, processing and analysis of multimedia data, as well as artificial and computational intelligence methods, have observed a close cooperation with each other during the last few years. The attempt has been to develop intelligent and efficient human computer interaction systems, enabling the user to access vast amounts of heterogeneous information, stored in different sites and archives. The focus of technological attempts has been on the analysis of digital video, due to its large amounts of spatiotemporal interrelations, which turns it into the most demanding and complex data structure. Video understanding and semantic information extraction have been identified as important steps towards more efficient manipulation of visual media. Attaching metadata to the original content is the means for achieving this goal.

Unquestionably, the problem of finding the information that user desires is still an open research problem, although maintaining large archives of digital media is increasingly easy and cost-effective, since cheap storage capacity and network bandwidth are available. The problems of *information overload* (the user is flooded with documents which are irrelevant to the query) and *mismatch* (a document which the user desires does not match the query), which are commonly encountered in textual retrieval systems [19][7] are more severe for multimedia retrieval, as the multimedia content is not in the same form as the query [11] [20]. Thus, a more efficient approach to content-based retrieval is necessary when referring to multimedia documents.

* This work has been partially funded by the EU IST-1999-20502 FAETHON project and by the "HRAKLEITOS" project. The Project "HRAKLEITOS" is co-funded by the European Social Fund (75%) and National Resources (25%).

M. Wallace et al.: *Automatic Thematic Categorization of Multimedia Documents using Ontological Information and Fuzzy Algebra*, StudFuzz 204, 247–274 (2006)
www.springerlink.com

Two main approaches have been utilized for the problem of multimedia information retrieval. The first one, *query by example*, replaces the textual query with an exemplar multimedia document. Visual features are extracted from the example and matched to similar features extracted from every document [10]. Although this approach avoids the problem of matching multimedia to textual terms, it has other difficulties: it is often difficult to find an example of the document one desires, and more importantly, it is not easy to isolate from the example the feature(s) that the user considers relevant to his or her desire. Additionally, in [9] and [8], hybrid methods extending the query by example strategy are developed. The second one, *semantic indexing* [13], retains the textual form of the query and attempts to extract, from the multimedia document, meaningful entities, such as objects and events, which can be mapped to textual terms and then matched to the textual query. This approach makes it possible to adopt successful expansion and adaptation techniques from the textual retrieval paradigm in multimedia retrieval.

Still, many extensions to research completed in textual information retrieval are needed. A semantic approach to content-based multimedia information retrieval will have to tackle at least the following issues:

- analyze metadata and raw multimedia data for the detection of higher level entities, i.e. for the determination of documents' semantics.
- analyze user requests, as to determine their semantic content.
- efficiently match the semantics of user requests to the semantics of documents.

In this chapter we focus on the first of these issues and analyze metadata towards detection of high level semantics. Specifically, starting from a multimedia document, a set of algorithms performing automated spatiotemporal segmentation is implemented. These objects and event detection techniques are integrated in a video annotator application. The user is free to manually select detected objects and events and input free text as annotation, constructing the required semantic index. The application then produces an MPEG-7 annotation that fully describes the hierarchical structure of the document and its semantics at each level of description. Consequently, utilizing semantic ontological information together with a novel definition of context that relies on fuzzy algebra, we estimate the semantic content of the document. The proposed methodology considers and tackles uncertainty in the generation of the semantic index, as well as fuzziness of real life semantics via ontological relations.

The structure of the chapter is as follows. In Sect. 2 we provide a description of the annotator application and the semantic index construction. In Sect. 3 we present the knowledge representation utilized by our approach, introducing a novel fuzzy quasi-taxonomic relation to model real life information. In Sect. 4 we then provide the methodology for detecting thematic categories within multimedia documents. Towards that scope, we utilize the notion of semantic context and propose a fuzzy hierarchical clustering algorithm for the extraction of topics. Finally, Sect. 5 provides experimental results and conclusions are drawn in Sect. 6.

2 Video Analysis, Annotation and Indexing

2.1 Multimedia Indexing

In the framework of the MPEG-7 standard [27], features of multimedia content are described through a rich set of data structures, termed Descriptors and Description Schemes. The former define the syntax and the semantics of the representation of features, while the latter specify the structure and semantics of the relationships between Descriptors or other Description Schemes. Low-level object features, such as color, shape, texture, and motion can be described via visual descriptors. Spatiotemporal relations can be described via the segment decomposition and relation tools. High-level features, such as objects, events and abstract concepts, are described via semantic description schemes such as the semantic entity. It is expected that, while most descriptors corresponding to low-level features are extracted automatically, human intervention will be required to produce the high-level descriptors.

Of course, there is no reason to assume that a document's description (either expressed in terms or in semantic entities) contains entities with equal importance. A more flexible and accurate approach is to weight the *importance* of the description terms, i.e. to attach a degree of importance to them. In text documents, a purely semantic degree of importance needs natural language understanding, which is not always feasible; simpler techniques to retrieve the weight of index terms include statistical analysis of terms [3]. Such techniques can also be applied on the textual annotation of multimedia documents. Moreover, in the case of annotation, weights can also be inferred from the underlying structure, e.g. terms found in the title could be weighted more than those found in the free text description. On the other hand, a multimedia document provides some cues that are useful in extracting the objects of interest. Such cues are motion (a person that moves is more important to a shot, than a table, for example), depth (an object in the foreground is in general more important than an object in the background) and context, if available (for example it is important to know if a shot is indoors or outdoors, whether it is in the sea or in a field).

A different source of weights is *uncertainty* [15]. Uncertainty in general plays little role in text indexing, and is measured with techniques such as stemming. Another source of uncertainty in text retrieval is context; for example the same word can have very different meanings depending on context. This is mainly due to the fact that text is not an encapsulated representation of an intended meaning, but rather is a cue that allows the anticipated audience to construct an appropriate meaning heavily based on context. On the other hand, uncertainty in computer vision problems is in general greater, as the descriptors and the models used to describe visual features are imperfect [22]. While importance and uncertainty are very different kinds of information [12], they are correlated, as an object considered to be important for the scene (such as a person in the foreground) is arguably recognized with greater certainty than one that is considered unimportant. Thus, we only use one number for weighting entities.

Still, the problem of semantic analysis of multimedia documents starting from raw multimedia information cannot be tackled in a fully automated manner, when even the problems of semantic segmentation and simple object detection remain open. Thus, we follow a manual approach: we utilize scene detection, object localization and tracking and feature detection techniques to prepare a syntactic description of multimedia content, which is then used by a human to provide textual annotation, at each level of description, using an annotation application and free text analysis.

Different video analysis techniques, such as temporal segmentation, object localization and tracking are integrated in a video annotator application [4]. A screenshot of its user interface is shown in Figure 1. This application is able to partition video into shots, cluster similar shots into higher level temporal segments, extract key frames, detect moving objects in a particular shot and let the user annotate the entire video program or each temporal segment, shot, key frame or object by associating it with keywords or free text. A hierarchical spatiotemporal video segmentation is thus achieved, and a textual MPEG-7 annotation is produced as output. This annotation fully describes a multimedia document at a syntactic level. The annotator is able to provide annotation at all levels, from the highest level of the video program or multimedia document to the lowest level of moving objects. This is achieved through the hierarchical structure of video segments, also depicted in Figure 1.

The main features supported by the annotator application are summarized below:

1. automatic shot detection, shot clustering and key frame extraction
2. object localization and tracking
3. hierarchical representation of multimedia segments (multimedia document → video sequence → theme → shot group → shot/key frame → moving object)
4. MPEG-7 compliant syntactic description and annotation

The annotator is free to choose the depth in which he decides to annotate. For example, he might annotate only the entire multimedia document or just part or all of the shots. Or he might not take the time to execute the automatic shot detection and limit himself to the annotation only. The overall structure and format of the annotation output is depicted in the MPEG-7 segment presented in Figure 2. This annotation consists of characteristic segments, emphasizing on the structure of the corresponding video content. The

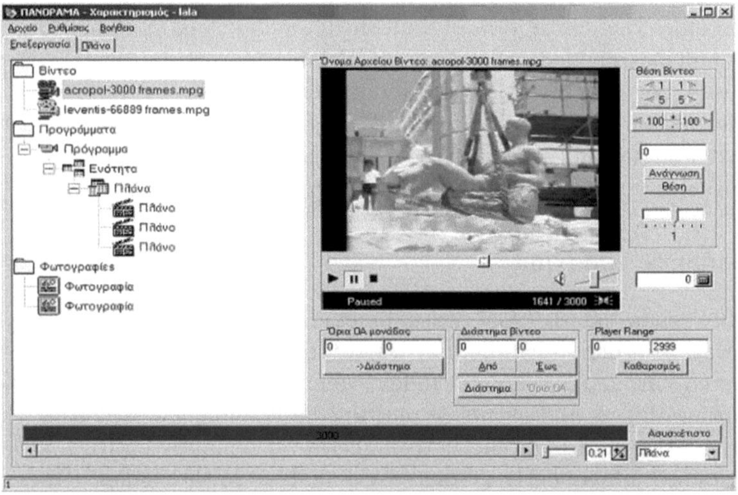

Fig. 1. User interface of the annotator application


```
<?xml version="1.0" encoding="ISO-8859-1"?>
<Mpeg7 xmlns="urn:mpeg:mpeg7:schema:2001" xmlns:xsi="http://www.w3.org/2001/XMLSchema-instance"
xsi:schemaLocation="mpeg7 mpeg\Mpeg7-2001.xsd">
    <DescriptionMetadata>
        <CreationTime>1975-10-18</CreationTime>
    </DescriptionMetadata>
    <Description xsi:type="ContentEntityType">
        <MultimediaContent xsi:type="AudioVisualType">
            <AudioVisual>
                <CreationInformation>
                    <Creation>
                        <Title>Konstantinos Karamanlis</Title>
                    </Creation>
                </CreationInformation>
                <TemporalDecomposition>
                    <AudioVisualSegment>
                        <MediaInformation>
                            <MediaProfile>
                                <MediaInstance>
                                    <InstanceIdentifier>1318</InstanceIdentifier>
                                    <MediaLocator>
                                        <MediaUri>Karamanlis.mpg</MediaUri>
                                    </MediaLocator>
                                </MediaInstance>
                            </MediaProfile>
                        </MediaInformation>
                        <CreationInformation>
                            <Creation>
                                <Title>Konstantinos Karamanlis talks to the people</Title>
                                <Abstract>
                                    <FreeTextAnnotation>Konstantinos Karamanlis addresses the crowd from a balcony. the crowd is
holding flags and shouts for his party. Shots of the crowd applausing and shouting.</FreeTextAnnotation>
                                    <KeywordAnnotation>
                                        <Keyword>Crowd</Keyword>
                                        <Keyword>Party</Keyword>
                                    </KeywordAnnotation>
                                </Abstract>
                            </Creation>
                        </CreationInformation>
                        <MediaTime>
                            <MediaTimePoint>T05:29:17</MediaTimePoint>
                            <MediaDuration>P0DT00H03M18S</MediaDuration>
                        </MediaTime>
                    </AudioVisualSegment>
                </TemporalDecomposition>
            </AudioVisual>
        </MultimediaContent>
    </Description>
</Mpeg7>
```


Fig. 2. MPEG-7 annotation of a segment of a multimedia document, produced by the annotator application

code provides the description of multimedia audiovisual content. The title of the content is provided, as well as significant information on the particular audiovisual segment: information about the media, its instance locator and uri, information about the creation of the fragment in the form of a title, a free text annotation and a set of keywords, as well as the duration of the audiovisual fragment.

2.2 Semantic Indexing

Manually generated MPEG-7 textual metadata need to be treated in a meaningful way, in order to handle user queries during information retrieval. This is achieved through a semantic index, which is automatically generated via recognizing objects and events in multimedia documents, exploiting the textual annotation and mapping it to higher-level semantic entities. Document semantics are subsequently extracted via analysis of the semantic index to be exploited in the retrieval process.

The index contains sets of document locators (links) for each semantic entity (thematic category, object, event, concept, agent, semantic place or semantic time), indicating in which document every entity is encountered. It has the form of a matrix, mapping documents to semantic entities or vice versa. Document locators associated to index entries may link to complete a/v documents, objects, still images or other video decomposition units that may be contained in the a/v archives.

For the automatic generation of the semantic index, semantic interpretation of the free text provided by experts as metadata using the annotator tool is performed, in order to match the textual terms to semantic entities. During this process, a query is issued for all documents that have not been indexed, or whose description has been updated. The textual annotation contained in the MPEG-7 compliant description of each such document is analyzed and semantic entities are identified through matching with their definitions in the knowledge model. Links between detected semantic entities and the document in question are then added to the index; weights are also added depending on the location of each entity in the description and the degree of entity's matching.

The definitions of semantic entities in the knowledge model may contain sequences of labels, each one providing a different textual form of the semantic entity, possibly in more than one language. Matching those to the terms in the textual annotation can provide the semantic representation of the annotation. Of course, in most cases this is far from trivial: the mapping between terms and semantic entities is a many-to-many relation, which means that multiple possible semantic interpretations exist for the same text.

As a simple example, consider the case of the term 'element'. At least two distinct semantic entities correspond to it: 'element1', which is related to chemistry, and 'element2', which is related to XML. Supposing that the annotation contains the term 'element', we need to determine which semantic entity the term should be mapped to. In the same example, if the remaining terms of the annotation are related to chemistry, then it is quite safe to suppose that the user is referring to semantic entity 'element1' rather than to semantic entity 'element2'. This implies that the context of the text can be used to facilitate the process of semantic entity determination in the case of ambiguities. However, the context can only be estimated using a representation of a fuzzy set of entities, and thus cannot be performed before the interpretation is completed. Consequently, interpretation needs to take place simultaneously with context estimation.

3 Knowledge Model for Semantic Document Analysis

Once the semantic index has been constructed, semantic document analysis and clustering is carried out in order to detect thematic categories in each multimedia document. This section provides the necessary background for the further analysis steps. In particular, we introduce a novel fuzzy quasi-taxonomic relation, constructed through a combination of suitable relations, in order to best handle knowledge and context representation.

The approach followed is based on a formal methodology that is founded on fuzzy relational algebra [14]. The mathematical notation used for this purpose is summarized in the following.

3.1 Ontological Relations

It is common knowledge that retrieval systems based on terms suffer from the problematic mapping of terms to concepts [18]. Specifically, as more than one terms may be associated to the same concept, and more than one concepts may be associated to the same term, the processing of query and index information is not trivial. In order to overcome such problems, we have chosen to work directly with concepts, rather than terms. We refer to these concepts as *semantic entities* [18]. In the sequel, we will denote by $S = \{s_1, s_2, \ldots, s_n\}$, the set of semantic entities that are known. The definitions of these entities, together with their textual descriptions, i.e. their corresponding terms, reside in a *semantic encyclopedia.* An encyclopedia additionally contains semantic relations amongst the semantic entities. Recently, much attention has been drawn to the design, representation, population and storage of such relations, using ontologies [16]. An ontology can be viewed as a framework for knowledge representation in which the context determines the intended meaning of each word. A word used in different context may have different meanings. In general, ontologies may be described as follows:

$$O = \{S, \{R_i\}\}, i = 1 \ldots n \tag{1}$$

$$R_i : S \times S \rightarrow \{0, 1\}, i = 1 \ldots n \tag{2}$$

where O is an ontology and R_i the i-th semantic relation amongst the semantic entities. The formal definition of ontologies also supports an inference layer, but this is outside the scope of this chapter.

Although any type of relation may be contained in an ontology, the two main categories are taxonomic (i.e. ordering) and compatibility (i.e. symmetric) relations. Compatibility relations have traditionally been exploited by information retrieval systems for tasks such as query expansion. They are ideal for the description of similarities of various natures, but fail to assist in the determination of the context of a query or a document; the use of ordering relations is necessary for such tasks [2]. Thus, a main challenge of intelligent multimedia retrieval is the meaningful exploitation of information contained in taxonomic relations of an ontology.

It is well understood that relations among real life entities are always a matter of degree, and are, therefore, best modelled using fuzzy relations. A rather self-explicable example is presented in Figure 3. Ontological taxonomies, on the other hand, are crisp in principle. Thus, they fail to fully describe real life concepts, and are limited to α-cuts of the desired relations. In order to explain the difference, consider A as a fuzzy subset of X, where X is a space of points or what is called a "Universe of Discourse". Fuzzy Set A in X is characterized by a membership function $\mu_A(x)$ which associates a real number in the interval $[0, 1]$ with each point x in X. Then its α-cut A_a is a non-fuzzy subset of X defined by $A_a = \{x|\mu_A(x) \geqq \alpha\}$. Figure 4 shows an α-cut on fuzzy set A.

This is a very important drawback, that makes such relations insufficient for the services that an intelligent multimedia retrieval system aims to offer.

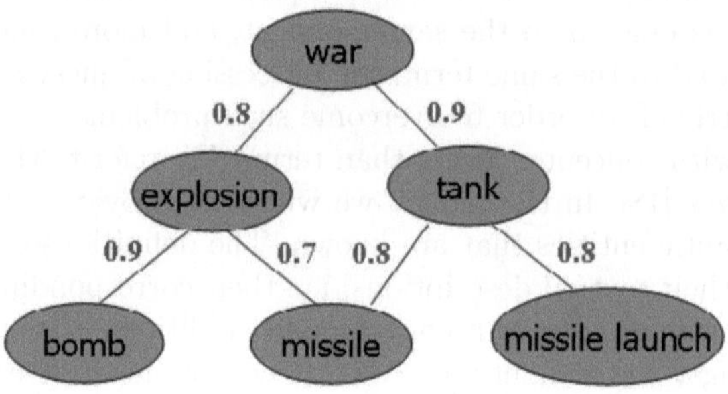

Fig. 3. Each relation is dominated by a degree d, which could be crisp (0 or 1) or fuzzy (i.e. 0.8)

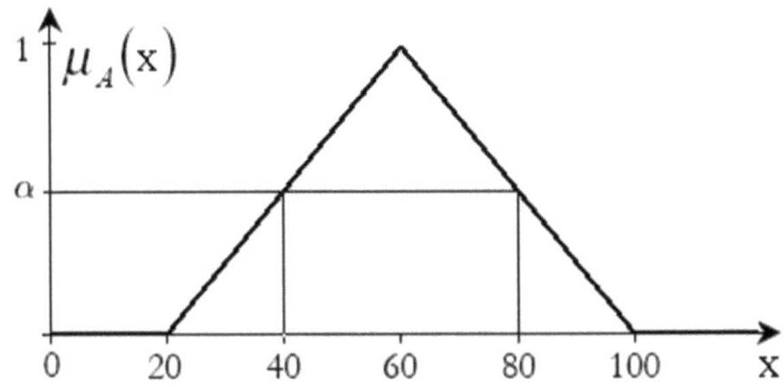

Fig. 4. α-cut on a fuzzy set A

3.2 The Fuzzy Quasi – Taxonomic Relation

In order to overcome such problems, fuzzy semantic relations have been proposed for the modelling of real life information [1]. In this section, we present a few commonly encountered semantic relations that can be modelled as fuzzy ordering relations, and propose their combination for the generation of a meaningful, fuzzy, quasi-taxonomic relation. Based on this relation, we will explain how the problem of automatic thematic categorization may be tackled in Sect. 4.

The *specialization* relation Sp is a fuzzy partial ordering relation on the set of semantic entities. $Sp(a, b) > 0$ means that the meaning of a "includes" the meaning of b; the most common form of specialization is sub-classing, i.e. a is a generalization of b. For example a could be a vehicle and b could be a car. The role of the specialization relation in knowledge-based retrieval is as follows: if a document refers to the meaning of entity b, then it is also related to a, since b is a special case of a. Still, there is no evidence that the opposite also holds; it is obvious that the specialization relation contains important information that cannot be modelled in a symmetric relation.

The *part* relation P is also a fuzzy partial ordering on the set of semantic entities. $P(a, b) > 0$ means that b is a part of a. For example a could be a human body and b could be a hand. The role of P in content-based retrieval is the opposite of that of Sp; if the user query contains b, then a document containing a will probably be of interest, because a contains a part b.

The *context* relation Ct is also a fuzzy partial ordering on the set of semantic entities. $Ct(a, b) > 0$ means that b provides the context for a or, in other words, that b is the thematic category that a belongs to. Other relations considered in the following have similar interpretations. Their names and corresponding notations are given in Table 1.

In this work, fuzziness of the aforementioned relations has the following meaning: High values of $Sp(a, b)$, imply that the meaning of b approaches the meaning of a, in the sense that when a document is related to b, then it is most probably related to a as well. On the other hand, as $Sp(a, b)$ decreases, the meaning of b becomes "narrower" than the meaning of a, in the sense that

Table 1. The fuzzy semantic relations

Sp	Specialization
Ct	Context
Ins	Instrument
P	Part
Pat	Patient
Loc	Location
Pr	Property

a document's relation to b will not imply a relation to a as well with a high probability, or to a high degree. Summarizing, the value of $Sp(a,b)$ indicates the degree to which the stored knowledge shows that an occurrence of b in a document implies relation to a. Likewise, the degrees of the other relations can also be interpreted as conditional probabilities or degrees of implied relevance.

The above imply that, for example, $a \neq b \rightarrow Sp(a,b) < 1$ since, if $a \neq b$, then we cannot be sure that both a and b are related to a given document, without first examining the document's context; at this point it is important to remind that a and b are not terms but concepts, which means that $a \neq b$ indicates / ensures a difference in a conceptual level.

A last point to consider is the transitivity of the relations presented above. It is obvious that if b is a specialization of a and c is a specialization of b, then c is a specialization of a. This implies that the specialization relation is transitive. A similar argument can be made for the other relations, as well. Let us now consider a more practical example. Let a be the concept of "car", b the concept of "wheel" and c the concept of "rubber". The inclusion $a < b < c$ is rather obvious. Still, it is not equally obvious that a user requesting documents related to rubber will be satisfied when faced with documents that are related to cars. By this example we wish to demonstrate that the form of transitivity used cannot be $max - min$ transitivity, but one relying on a subidempotent norm. Therefore, we demand that the presented relations are sup-t transitive, where t is an Archimedean norm. This means that $Sp(a,c) \geq \max_{s \in S} t(Sp(a,s), Sp(s,c))$, $t(a,a) < a$ and, therefore, $t(a,b) < \min(a,b), \forall a \in (0,1)$.

More formally, the knowledge model presented above may be summarized in the following:

$$O_{\mathcal{F}} = \{S, \{r_i\}\}, i = 1 \ldots n \tag{3}$$

$$r_i = \mathcal{F}(R_i) : S \times S \rightarrow [0,1], i = 1 \ldots n \tag{4}$$

The existence of many relations has lead to the need for utilization of more relations for the generation of an adequate taxonomic relation T. Based on the relations r_i we construct the following semantic relation:

$$T = Tr^t(\bigcup_i r_i^{p_i}), p_i \in \{-1, 1\}, i \in 1 \dots n \tag{5}$$

where $Tr^t(A)$ is the sup-t transitive closure of relation A; the transitivity of relation T was not implied by the definition, as the union of transitive relations is not necessarily transitive. In this work we use a taxonomic relation that has been generated with the use of the following semantic relations:

- Specialization *Sp*.
- Context *Ct*, inverted.
- Part *P*.
- Instrument *Ins*. $Ins(a, b) > 0$ indicates that b is an instrument of a. For example, a may be "music" and b may be "drums".
- Location *Loc*. $Loc(a, b) > 0$ indicates that b is the location of a. For example, a may be "concert" and b may be "stage".
- Patient *Pat*. $Pat(a, b) > 0$ indicates that b is a patient of a. For example, a may be "course" and b may be "student".
- Property *Pr*. $Pr(a, b) > 0$ indicates that b is a property of a. For example, a may be "Jordan" and b may be "star".

Thus, the utilized relation is:

$$T = Tr^t(Sp \cup Ct^{-1} \cup Ins \cup P \cup Pat \cup Loc \cup Pr) \tag{6}$$

An example of the T relation taken from the airplane domain is presented in Figure 5. Based on the semantics of relations r_i, it is easy to see that T is ideal for the determination of the thematic categories that an entity may be related to, as thematic categories are also semantic entities:

$$C \subseteq S \tag{7}$$

where $C = \{c_i\}, i \in 1 \dots k$ is the set of thematic categories (for example *ball* and *stadium* may be semantic entities, while *football* and *sports* are both semantic entities and thematic categories).

All the relations used for the generation of T are partial ordering relations. Still, there is no evidence that their union is also antisymmetric. Quite the contrary, T may vary from being a partial ordering to being an equivalence relation. This is an important observation, as true semantic relations also fit in this range (total symmetricity as well as total antisymmetricity often have to be abandoned when modelling real life). Still, the semantics of the used relations, as well as our experiments, indicate that T is "almost" antisymmetric. Therefore, we categorize to it as quasi-ordering or quasi-taxonomic.

4 Detection of Thematic Categories

In this section we focus on the extraction of semantic content of multimedia documents, in the form of thematic categorization. Specifically, we present initially the main problem formulation, whose aim is to detect which semantic

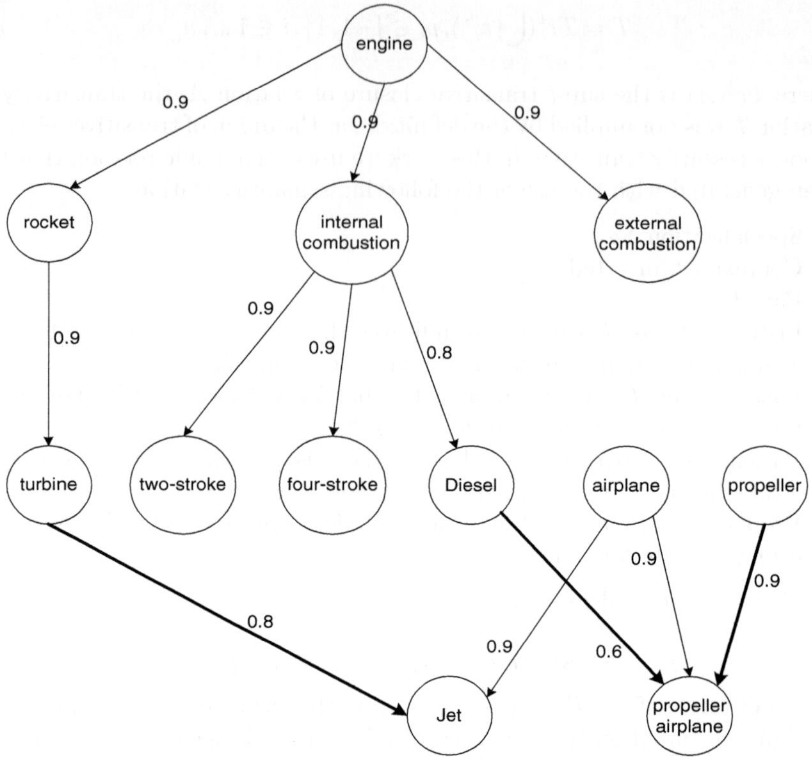

Fig. 5. Example of T relation construction

entities and thematic categories are related to a particular document. In the following, we explain how the notion of context can be defined using the aforementioned fuzzy quasi-taxonomic relation. Continuing, we explain how this context can be utilized to detect the thematic categories to which a document is related, while at the same time overcoming the problem of uncertainty or noise in the semantic index, based on a fuzzy hierarchical clustering approach.

4.1 Problem Formulation

Let us first present the problem that this work attempts to address, in a more formal manner. The main objective is to analyze the semantic index, with the aim of extracting a document's semantics. In other words, we aim to detect which semantic entities and thematic categories are indeed related to a document, and to which extent. More formally, we accept as input the semantic indexing of available documents, i.e. the semantic index I. This is in fact a fuzzy relation between the sets of documents D and semantic entities S:

$$I : D \times S \rightarrow [0, 1] \tag{8}$$

Each document d is represented as a normal fuzzy set $I(d)$ on the set of semantic entities, i.e.:

$$\forall d \in D \;\; \exists s \in S \text{ such that } I(s, d) = 1$$

Based on this set, and the knowledge contained in the available semantic relations, we aim to detect the degree to which a given document d is related to a semantic entity $s \in S$. This entity may be (and usually is) new with respect to set $I(d)$, i.e. it may not be already known to be associated with document d simply based on the document indexing process. We will refer to this degree as $R_T(s, d)$. In other words, we attempt to calculate a relation

$$R_T : S \times D \rightarrow [0, 1] \tag{9}$$

where D is the set of available documents, as already explained. In designing an algorithm that is able to calculate this relation in a meaningful manner, a series of issues need to be tackled. Examples of such issues are depicted in Figure 6.

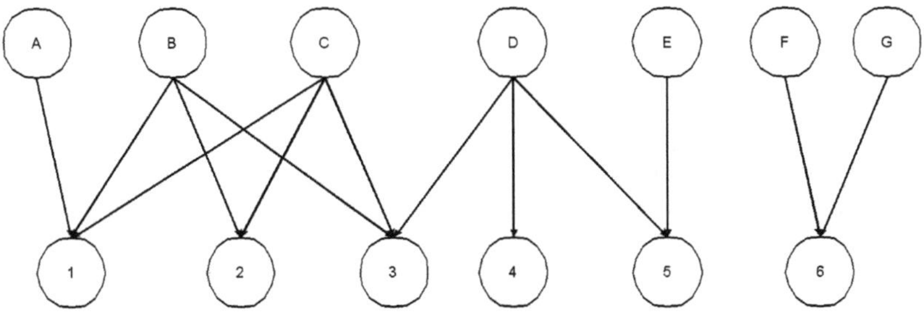

Fig. 6. Entities that index a document (1,2,..,6), related topics detected (A,B,...,G) and relations among them.

Among others we observe that:

- there are no topics that relate all the entities
- entity 1 is related to topics A, B and C, but only the latter two are related to the whole document, due to their numerous (> 1) relations to distinct topics within the whole document. Also, topic A is not necessarily related to the whole document.
- entity 3 is related to two distinct topics of interest. Topics B and C are considered a team, since they relate to the exact same entities
- topics F, G are related to only one of the document's entities; this could be coincidental.

Consequently, the above example illustrates all the issues to be tackled in designing an efficient algorithm, which can be summarized into the following:

1. A semantic entity may be related to multiple, unrelated topics. Example: a ball may be related to baseball, basketball, kids ball, etc. Consequently, the common meaning of the remaining entities that index the given document has to be considered.
2. A document may be related to multiple, unrelated topics. Example: fans in a stadium may imply football match, concert, protest, etc. If this is the case, most entities will be related to just one of these topics. Therefore, clustering of the remaining entities, based on their common meaning, needs to be applied.
3. The semantic index may contain incorrectly recognized entities. Example: entities from the use of terms in metaphorical sense. Those entities will not be found similar to other entities, so the cardinality of the clusters can be used.

In the following, keeping these issues in mind, we provide the principles of the applied approach. According to issue (1), a semantic entity may correspond to multiple, unrelated topics. Therefore, it is necessary for the algorithm to be able to determine which of these topics are indeed related to a given document. In order for this task to be performed in a meaningful manner, the common meaning of the remaining entities that index the given document needs to be considered as well.

On the other hand, when a document is related to more than one, unrelated topics, as issue (2) points out, we should not expect all the entities that index it to be related to each one of the topics in question. Quite the contrary, we should expect most entities to be related to just one of these them. Therefore, a clustering of semantic entities, based on their common meaning, needs to be applied.

In this process, entities that are misleading (e.g. entities that resulted from incorrect detection of entities in the document) will probably not be found similar with other entities that index the given document. Therefore, the cardinality of the clusters may be used to tackle issue (3).

The proposed approach may be decomposed into the following steps:

1. Create a single taxonomic semantic relation that is suitable for use by the thematic categorization module.
2. Determine the count of distinct topics that a document is related to, by performing a partitioning of semantic entities, using their common meaning as clustering criterion.
3. Fuzzify the partitioning, in order to allow for overlapping of clusters and fuzzy membership degrees.
4. Identify the topic that is related to each cluster.
5. Aggregate the topics for distinct clusters in order to acquire an overall result for the document.

Each of the above steps uses the taxonomy relation, in addition to the index. In the following, after discussing the notion of "common meaning", we elaborate on each of these steps.

4.2 The Notion of Context

We have shown that in the process of content analysis we have to use the common meaning of semantic entities. We will refer to this as their *context* [2]; in general, the term context refers to whatever is common among a set of elements. Relation T will be used for the detection the context of a set of semantic entities, as explained in the remaining of this subsection.

A document d is represented only by its mapping to semantic entities, via the semantic index. Therefore, the context of a document is again defined via the semantic entities that are related to it. The fact that relation T described in subsection 3.2 is (almost) an ordering relation allows us to use it in order to define, extract and use the context of a document, or a set of semantic entities in general.

Relying on the semantics of relation T, we define the context $K(s)$ of a single semantic entity $s \in S$ as the set of its antecedents in relation T. More formally, $K(s) = T(s)$. Assuming that a set of entities $A \subseteq S$ is crisp, i.e. all considered entities belong to the set with degree one, the context of the group, which is again a set of semantic entities, can be defined simply as the set of their common antecedents:

$$K(A) = \bigcap_i K(s_i), s_i \in A \tag{10}$$

Obviously, as more entities are considered, the context becomes narrower, i.e. it contains less entities and to smaller degrees (Figure 7):

$$A \supset B \rightarrow K(A) \subseteq K(B) \tag{11}$$

When the definition of context is extended to the case of fuzzy sets of semantic entities, i.e., A is fuzzy, this property must still hold. Moreover, we demand that the following are satisfied as well, basically because of the nature of fuzzy sets:

- $A(s) = 0 \implies K(A) = K(A - \{s\})$, i.e. no narrowing of context.
- $A(s) = 1 \implies K(A) \subseteq K(s)$, i.e. full narrowing of context.
- $K(A)$ decreases monotonically with respect to $A(s)$.

Taking these into consideration, we demand that, when A is a normal fuzzy set, the "considered" context $\mathcal{K}(s)$ of s, i.e. the entity's context when taking its degree of participation to the set into account, is low when the degree of participation $A(s)$ is high, or when the context of the crisp entity $K(s)$ is low. Therefore

$$\mathcal{K}(s) \doteq K(s) \cup cp(S \cdot A(s)) \tag{12}$$

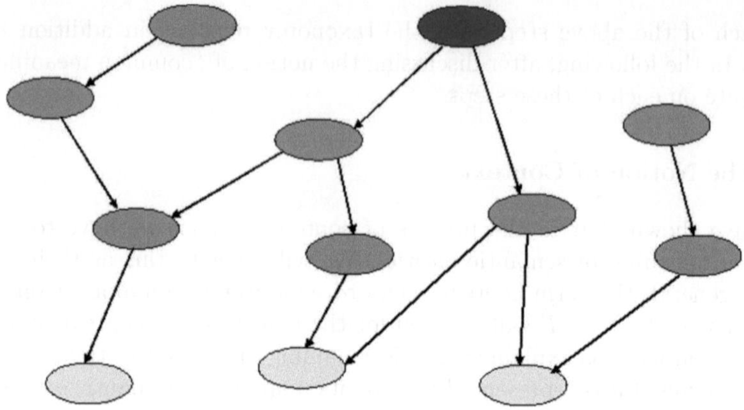

Fig. 7. As more entities are considered, the context it contains less entities and to smaller degrees: Considering only the first two leaves from the left, the context contains two entities, whereas considering all the leaves narrows the context to just one common descendant

where *cp* is an involutive fuzzy complement and $S \cdot A(s)$ is a fuzzy set for which

$$[S \cdot A(s)]_{(x)} = A(s) \forall x \in S \tag{13}$$

Then the set's context is easily calculated as

$$K(A) = \bigcap_i K(s_i), s_i \in A \tag{14}$$

Considering the semantics of the T relation and the process of context determination, it is easy to realize that when the entities in a set are highly related to a common meaning, the context will have high degrees of membership for the entities that represent this common meaning. Therefore, we introduce the height of the context $h(K(A))$, which may be used as a measure of the semantic correlation of entities in fuzzy set A. We will refer to this measure as *intensity* of the context. The intensity of the context demonstrates the degree of relevance, as shown in Figure 8.

4.3 Fuzzy Hierarchical Clustering and Topic Extraction

Before detecting the topics that are related to a document d and in order to support the possibility of existence of multiple distinct topics in a single document, the set of semantic entities that are related to it needs to be clustered, according to their common meaning. More specifically, the set to be clustered is the support of the document:

$$^{0+}d = \{s \in S : I(s,d) > 0\} \tag{15}$$

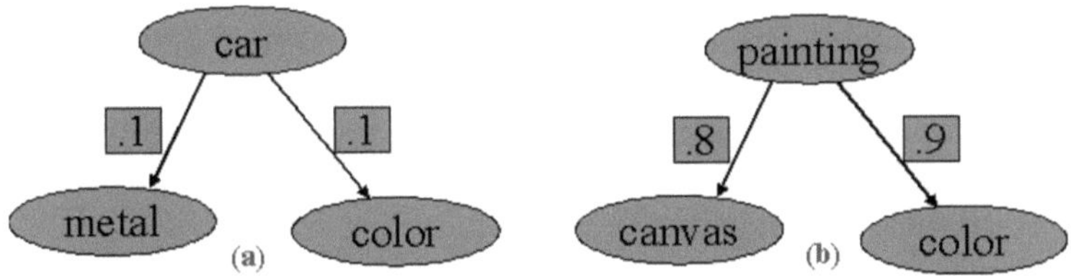

Fig. 8. Examples of different heights of context: (a) In the first set of entities, the degree of relevance is rather small and equal, so the height of the context is also small. (b) The second set of entities presents a differentiation in the degree of relevance between the two entities, so the height of the context is greater

Most clustering methods belong to either of two general categories, partitioning and hierarchical [21]. Partitioning methods create a crisp or fuzzy clustering of a given data set, but require the number of clusters as input. Since the number of topics that exist in a document is not known beforehand, partitioning methods are inapplicable for the task at hand [17]; a hierarchical clustering algorithm needs to be applied.

Hierarchical methods are divided into agglomerative and divisive. Of those, the first are more widely studied and applied, as well as more robust. Their general structure, adjusted for the needs of the problem at hand, is as follows:

1. When considering document d, turn each semantic entity $s \in^{0+} d$ into a singleton, i.e. into a cluster c of its own.
2. For each pair of clusters c_1, c_2 calculate a compatibility indicator $CI(c_1, c_2)$. The CI is also referred to as cluster similarity, or distance metric.
3. Merge the pair of clusters that have the best CI. Depending on whether this is a similarity or a distance metric, the best indicator could be selected using the max or the min operator, respectively.
4. Continue at step 2, until the termination criterion is satisfied. The termination criterion most commonly used is the definition of a threshold for the value of the best compatibility indicator.

The two key points in hierarchical clustering are the identification of the clusters to merge at each step, i.e. the definition of a meaningful metric for CI, and the identification of the optimal terminating step, i.e. the definition of a meaningful termination criterion.

When clustering semantic entities, the ideal distance metric for two clusters c_1, c_2 is one that quantifies their semantic correlation. In the previous subsection we have defined such a metric, the intensity of their common context $h(K(c_1 \cup c_2))$. Therefore, the process of merging of clusters will be based on this measure and should terminate when the entities are clustered into sets that correspond to distinct topics. We may identify such sets by the fact that their common contexts will have low intensity. Therefore, the termination criterion shall be a threshold on the selected compatibility metric.

Hierarchical clustering methods are more flexible than their partitioning counterparts, in that they do not need the number of clusters as an input. This clustering method, being a hierarchical one, will successfully determine the count of distinct clusters that exist in ${}^{0+}d$. Still, it is less robust and inferior to partitioning approaches in the following senses:

- It only creates crisp clusters, i.e. it does not allow for degrees of membership in the output.
- It only creates partitions, i.e. it does not allow for overlapping among the detected clusters.

Both of the above are great disadvantages for the problem at hand, as they are not compatible with the task's semantics: in real life, a semantic entity may be related to a topic to a degree other than 1 or 0, and may also be related to more than one distinct topics. In order to overcome such problems, we describe in the following a method for fuzzification of the partitioning. Thus, the clusters' scalar cardinalities will be corrected, so that they may be used later on for the filtering of misleading entities.

Each cluster is described by the crisp set of semantic entities $c \subseteq^{0+} d$ that belong to it. Using those, we may construct a fuzzy classifier, i.e. a function C_c that measures the degree of correlation of a semantic entity s with cluster c.

$$C_c : S \to [0, 1]$$

Obviously, a semantic entity s should be considered correlated with c, if it is related to the common meaning of the semantic entities in c. Therefore, the quantity

$$C_1(c, s) = h(K(c \cup \{s\})) \tag{16}$$

where $h(\cdot)$ symbolizes the height of a fuzzy set, is a meaningful measure of correlation. Of course, not all clusters are equally compact; we may measure cluster compactness using the similarity among the entities it contains, i.e. using the intensity of the cluster's context. Therefore, the aforementioned correlation measure needs to be adjusted, to the characteristics of the cluster in question:

$$C_2(c, s) = \frac{C_1(c, s)}{h(K(c))}$$

It is easy to see that this measure obviously has the following properties:

- $C_2(c, s) = 1$ if the semantics of s imply it should belong to c. For example $C_2(c, s) = 1, \forall s \in c$
- $C_2(c, s) = 0$ if the semantics of s imply it should not belong to c.
- $C_2(c, s) \in (0, 1)$ if s is neither totally related, nor totally unrelated to c.

These are the properties that we wish for the cluster's fuzzy classifier, so we define the correlation of s with c as:

$$C_c(s) \doteq C_2(c,s) = \frac{C_1(c,s)}{h(K(c))} = \frac{h(K(c \cup \{s\}))}{h(K(c))} \tag{17}$$

Using such classifiers, we may expand the detected crisp partitions to include more semantic entities. Cluster c is replaced by the fuzzy cluster c^f:

$$c^f = \sum_{s \in {}^{0+}d} s/C_c(s) \tag{18}$$

Obviously $c^f \supseteq c$.

The process of fuzzy hierarchical clustering has been based on the crisp set ${}^{0+}d$, thus ignoring fuzziness in the semantic index. In order to incorporate this information when calculating the clusters that describe a document's content, we adjust the degrees of membership for them as follows:

$$c^i(s) = t(c^f(s), I(s,d)), \forall s \in {}^{0+}d \tag{19}$$

where t is a t-norm. The semantic nature of this operation demands that t is an Archimedean norm. Each one of the resulting clusters corresponds to one of the distinct topics of the document.

In order to determine the topics that are related to a cluster c^i, two things need to be considered: the scalar cardinality of the cluster $|c^i|$ and its context. Since context has been defined only for normal fuzzy sets, we need to first normalize the cluster as follows:

$$c^n(s) = \frac{c^i(s)}{h(c^i(s))}, \forall s \in {}^{0+}d \tag{20}$$

Obviously, semantic entities that are not contained in the context of c^n cannot be considered as being related to the topic of the cluster. Therefore

$$R_T(c^i) \subseteq R_T^*(c^n) = w(K(c^n)) \tag{21}$$

where w is a weak modifier. Modifiers, which are also met in the literature as *linguistic hedges* [14], are used in this work to adjust mathematically computed values so as to match their semantically anticipated counterparts.

In the case where the semantic entities that index document d are all clustered in a unique cluster c^i, then $R_T(d) = R_T^*(c^n)$ is a meaningful approach, where R_T^* corresponds to the output in case of neglecting cluster cardinality. On the other hand, when more than one clusters are detected, then it is imperative that cluster cardinalities are considered as well.

Clusters of extremely low cardinality probably only contain misleading entities, and therefore need to be ignored in the estimation of $R_T(d)$. On the contrary, clusters of high cardinality almost certainly correspond to the distinct topics that d is related to, and need to be considered in the estimation of $R_T(d)$. The notion of "high cardinality" is modelled with the use of a "large" fuzzy number $L(\cdot)$, which forms a function from the set of real positive numbers to the $[0,1]$ interval, quantifying the notion of "large" or "high". Accordingly, $L(a)$ is the truth value of the proposition "a is high", and, consequently, $L(|b|)$ is the truth value of the proposition "the cardinality of cluster b is high".

The topics that are related to each cluster are computed, after adjusting membership degrees according to scalar cardinalities, as follows:

$$R_T(c^i) = R_T^*(c^n) \cdot L(|c^i|) \tag{22}$$

The set of topics that correspond to a document is the set of topics that belong to any of the detected clusters of semantic entities that index the given document.

$$R_T(d) = \bigcup_{c^i \in G} R_T(c^i) \tag{23}$$

where $\bigcup$ is a fuzzy co-norm and G is the set of fuzzy clusters that have been detected in d. It is easy to see that $R_T(s,d)$ will be high if a cluster c^i, whose context contains s, is detected in d, and additionally, the cardinality of c^i is high and the degree of membership of s in the context of the cluster is also high (i.e., if the topic is related to the cluster and the cluster does not comprised of misleading entities).

5 Examples and Results

A first experiment for the validation of the proposed methodology is presented in the sequel, involving the thematic categorization of five multimedia documents. The documents have been processed and manually annotated using the tools presented in Sect. 2. A limited set of semantic entities have been automatically extracted from the textual annotation to construct the semantic index.

The semantic entities included in the manually constructed, limited knowledge base for this purpose are shown in Table 2, with thematic categories shown in boldface. The taxonomy relation available is shown in Table 3, using the entity mnemonics of Table 2. Zero elements of the relations, as well as elements that are implied by reflexivity are omitted. A portion of the semantic index constructed for the five documents is shown in Table 4, where the entities detected in document d_5 are omitted from the table and presented in the text below. Finally, the results of the algorithm for the detection of thematic categories in the documents are shown in Table 5.

Document d_1 contains a shot of a theater hall. The play is related to war. We can see that objects and events are detected with a limited degree of

Table 2. Semantic Entity names

S.Entity	*Mnemonic*	*S.Entity*	*Mnemonic*
arts	**art**	army or police uniform	unf
tank	tnk	lawn	lwn
missile	msl	goal	gol
scene	scn	shoot	sht
war	**war**	tier	tir
cinema	**cnm**	river	riv
performer	prf	speak	spk
sitting person	spr	F16	f16
explosion	exp	football player	fpl
launch of missile	lms	goalkeeper	glk
screen	scr	**theater**	**thr**
football	**fbl**	fighter airplane	far
curtain	crn	seat	sit

Table 3. The taxonomy relation

s_1	s_2	$T(s_1,s_2)$	s_1	s_2	$T(s_1,s_2)$	s_1	s_2	$T(s_1,s_2)$
war	unf	0.90	war	exp	0.60	war	lms	0.70
war	far	0.80	fbl	gol	0.80	fbl	lwn	0.90
war	tnk	0.80	fbl	sit	0.60	cnm	scr	0.90
war	msl	0.80	cnm	sit	0.60	cnm	spr	0.80
thr	scn	0.90	fbl	sht	0.90	fbl	spr	0.60
thr	prf	0.90	fbl	tir	0.80	thr	sit	0.60
thr	spr	0.80	fbl	fpl	0.90	thr	crn	0.70
far	f16	1.00	art	cnm	0.80	art	thr	0.80
fpl	glk	1.00						

Table 4. The Semantic Index

s	$d_1(s)$	s	$d_2(s)$	s	$d_3(s)$	s	$d_4(s)$
prf	0,9	spr	0.9	spr	0.8	spr	0.2
spr	0.9	spk	0.8	unf	0.9	unf	0.3
spk	0.6	sit	0.9	lwn	0.6	lwn	0.4
sit	0.7	scr	1.00	gol	0.9	gol	0.3
crn	0.8	tnk	0.4	tir	0.7	tir	0.4
scn	0.9			spk	0.9	spk	0.2
tnk	0.7			glk	0.6	glk	0.3
				sht	0.5	sht	0.4

Table 5. The result of semantic document analysis

	$R_T(d_1)$	$R_T(d_2)$	$R_T(d_3)$	$R_T(d_4)$	$R_T(d_5)$
arts	0.84	0.73			0.85
cinema		0.74			0.86
theater	0.89				0.33
football			0.84	0.37	0.77
war					0.77

certainty. Furthermore, detected entities are not always directly related to the overall topic of the document (for example a "tank" may appear in a shot from a theater, as a part of the play, but this is not a piece of information that can aid in the process of thematic categorization). The algorithm of document analysis ignores "tank" and "speak".

Document d_2 contains a shot from a cinema hall. The film is again related to war. Although some entities are common between d_1 and d_2 (and they are related to both "theater" and "cinema"), the algorithm correctly detects that in this case the overall topic is different. This is accomplished by considering that "screen" alters the context and thus the overall meaning.

Documents d_3 and d_4 are both related to football. Their difference is the certainty with which entities have been detected in them. As can be seen, the algorithm successfully incorporates uncertainty of the input in its result.

As a last example, document d_5 is a sequence of shots from a news broadcast. Due to the diversity of stories presented in it, the semantic entities that are detected and included in the index are quite unrelated to each other. Using the sum notation for fuzzy sets,

d_5= spr/0.9 + unf/0.8 + lwn/0.5 + gol/0.9 + tir/0.7 + spk/0.9 + glk/0.8 + sht/0.5 + prf/0.7 + sit/0.9 + crn/0.7 + scn/0.8 + tnk/0.9 + msl/0.8 + exp/0.9 + riv/1

After the consideration of the fuzziness of the index, the following five fuzzy clusters of entities are created:

c_1= spk/0.9

c_2= riv/1.0

c_3= spr/0.9 + prf/0.7 + sit/0.77 + crn/0.7 + scn/0.8

c_4= spr/0.9 + lwn/0.5 + gol/0.9 + tir/0.7 + glk/0.8 + sht/0.5 + sit/0.9

c_5= unf/0.8 + tnk/0.9 + msl/0.8 + exp/0.9

We can observe that the algorithm successfully identifies the existence of more than one distinct topics in the document. Furthermore, entities such as "seat" and "sitting-person" are assigned to more than one clusters, as they are related to more than one of the contexts that are detected in the document. In the following steps of the algorithm, the first two clusters are ignored, due to their small scalar cardinality.

The methodology described so far has been used in the design and the implementation of the Detection of Thematic Categories (DTC) module, an internal intelligent module of the Faethon multimedia mediator system [6]. In the Faethon system, the role of the DTC module is to parse document annotations and provide thematic categorization for them; this is then used in order to facilitate browsing, searching and personalization tasks. The mediator system integrates five archives, different in architecture, content and annotation language [23]. These are ERT (the Hellenic Public Broadcasting Corporation), FAA (Film Archive Austria), Alinari Archive (Italy), ORF (Austria) and FAG (Film Archive Greece). In the working prototype of the system each archive participates with approximately 200 documents, resulting in a total number of 1005 annotated multimedia documents [24]. WordNet synsets have been used as a source for the definition of the core body of the knowledge base semantic entities, resulting in over 70000 semantic entities.

The list of semantic entities that are characterized as thematic categories appears in the first column of Table 6. The second through sixth columns of Table 6 present the count of documents from each archive that match each thematic category; as the estimated relevance of documents to thematic categories using the methodology of this chapter is a matter of degree, a threshold of $T_c = 0.6$ is used in order to acquire crisp estimations of thematic categorization. We can see that, although some archives (e.g. FAA) have more documents related to sports while others (e.g. ERT) more related to military issues, otherwise all archives contain documents related to most thematic categories. The last column of the table presents the total count of documents (considering all five archives) that are related to each thematic category. It can be seen that documents map to the thematic categories in a rather uniform way, which makes the thematic categorization a powerful tool for retrieval and personalization tasks [24], [25].

In Table 7 we present the distribution of documents to thematic categories, i.e., the count of documents that are not related to any thematic categories, are related to exactly one thematic category, to exactly two thematic categories and so on. The majority of documents are related to multiple thematic categories - typically from 4 to 8 - which validates our fuzzy clustering approach; without it, classification of a document to multiple thematic categories would not have been possible.

In order to evaluate the accuracy and validity of the thematic categorization results, a precision-recall diagram has been constructed. In information retrieval (IR), precision is defined as the number of retrieved relevant items over the number of total retrieved items. Recall is defined as the number of retrieved relevant items over the total number of relevant items:

$$p = precision = \frac{relevant\ retrieved\ items}{retrieved\ items} \tag{24}$$

$$r = recall = \frac{relevant\ retrieved\ items}{relevant\ items} \tag{25}$$

Table 6. Thematic categories and archives

	ERT	*FAA*	*Alinari*	*ORF*	*FAG*	*Total*
business	14	6	54	39	6	119
history	78	26	23	24	195	346
olympics	12	87	45	16	14	174
football	5	181	144	68	68	466
sports	19	127	96	40	43	325
basketball	7	96	168	77	79	427
news	111	32	124	164	123	554
military	176	5	80	101	165	527
swimming	24	58	64	24	36	206
tennis	23	114	36	97	12	282
theater	98	2	67	69	45	281
politics	126	15	22	95	133	391
arts	135	13	96	74	91	409
commerce	35	26	23	11	17	112
technology	43	32	5	64	3	147
entertainment	55	45	175	137	74	486
health	63	12	34	42	35	186
education	23	7	76	32	2	140
music	78	33	36	92	66	305
cinema	98	24	165	149	52	488
nature	34	4	74	46	10	168
science	9	26	55	60	16	166
war	139	14	88	121	87	449

Table 7. Distribution of documents to thematic categories

Categories	*None*	1	2	3	4	5	6	7	8	9	10+
Documents	23	47	73	123	136	145	161	111	83	61	42

In our case, of course, "retrieval" refers to thematic categorization. Alternative terminologies are also widely used in classification problems, e.g. [26]. The performance for an "deal" system is to have both high precision and recall. Unfortunately, these are conflicting entities and cannot be at high values at the same time. Therefore, instead of using a single value of precision and recall, a Precision-Recall (PR) graph is typically used to characterize the performance of an IR system.

In order to acquire multiple pairs of precision and recall and draw the diagram, different thresholds have been employed on the degree of relevance of a document to a thematic category, i.e., threshold T_c was let vary from 0.3 to 0.9. Binary labels were manually assigned to all 1005 documents for five thematic categories (sports, military, arts, education and entertainment), in order to construct a ground-truth to be used for comparisons to the results of thematic categorization and subsequent measurement of precision and recall. The resulting diagram is presented in Figure 9 in red.

The yellow line of the same figure presents the precision-recall diagram for the Faethon system when the same five thematic categories are used as queries and thematic categorization information is not used in the query processing. We note that the Faethon query processing scheme also takes advantage of the knowledge stored in the encyclopedia and considers the context in query interpretation, query expansion and index matching operations [2]. Thus, any difference in the two diagrams reflects the operation of the proposed thematic categorization algorithm. We can see that for similar values of recall the thematic categorization has higher precision values, as it does not include documents that contain related words but are not truly related to the thematic category.

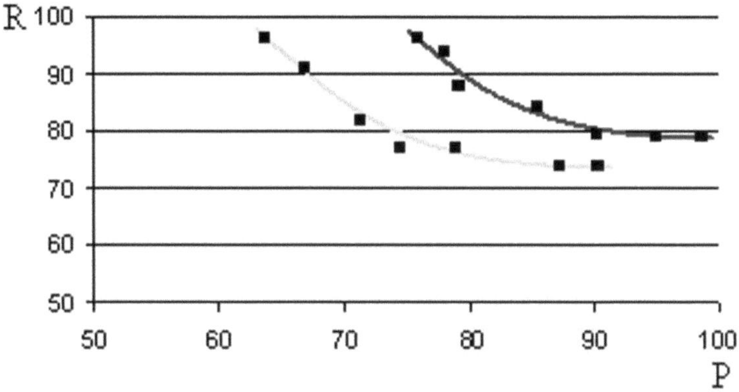

Fig. 9. Precision-Recall Diagram

6 Conclusions

The semantic gap refers to the inability to efficiently match document semantics with user semantics, mainly because neither is usually readily available in a useful form. In this work we have made an attempt to extract the former, relying on fuzzy algebra and a knowledge base of fuzzy semantic relations.

Specifically, we started by describing the construction of a semantic index using a hybrid approach of processing and manual annotation of raw multimedia information. We then explained how the index can be analyzed for the detection of the topics that are related to each multimedia document. The existence of noise and uncertainty in the semantic index has also been considered. Our approach is based on the notion of context and the utilization of fuzzy taxonomic relations.

As multimedia content is becoming a major part of more and more applications every day, the applications of this work are numerous. As more important, one may mention automated multimedia content organization, indexing and retrieval, usage history analysis, user adaptation, efficient multimedia content filtering and semantic unification of diverse audiovisual archives [5].

Although this work is contributing in the direction of bridging the semantic gap, a lot more has to be done before one may claim that the problem is solved. In this chapter we have assumed the existence of a semantic index, which cannot yet be constructed in an automated manner. Major focus in our future work will be given to the automated mapping of MPEG-7 syntactically described objects and events to their corresponding semantic entities, based on techniques such as graph matching. Another area of future research is the selection of optimal fuzzy operators for the most meaningful semantic output. Our findings so far indicate that this selection is not independent from the knowledge itself. Finally, one more direction is the utilization of existing crisp taxonomies for the generation of the knowledge that is required for the analysis of the multimedia documents.

References

1. Akrivas G., Stamou G., Fuzzy Semantic Association of Audiovisual Document Descriptions, *Proceedings of International Workshop on Very Low Bitrate Video Coding (VLBV)*, Athens, Greece, Oct. 2001.
2. Akrivas G., Wallace M., Andreou G., Stamou G. and Kollias S., Context – Sensitive Semantic Query Expansion, *Proceedings of the IEEE International Conference on Artificial Intelligence Systems (ICAIS)*, Divnomorskoe, Russia, September 2002.
3. Akrivas G., Wallace M., Stamou G. and Kollias S., Context – Sensitive Query Expansion Based on Fuzzy Clustering of Index Terms, *Proceedings of the Fifth International Conference on Flexible Query Answering Systems (FQAS)*, Copenhagen, Denmark, October 2002.
4. Akrivas G., Ioannou S., Karakoulakis E., Karpouzis K., Avrithis Y., Delopoulos A., Kollias S., Varlamis I. and Vaziriannis M., An Intelligent System for Retrieval and Mining of Audiovisual Material Based on the MPEG-7 Description Schemes, *Proceedings of European Symposium on Intelligent Technologies, Hybrid Systems and their Implementation on Smart Adaptive Systems (EUNITE 01)*, Tenerife, Spain, December 12-14, 2001.
5. Avrithis Y. and Stamou G., FAETHON: Unified Intelligent Access to Heterogenous Audiovisual Content, *Proceedings of the International Workshop on Very Low Bitrate Video Coding (VLBV)*, Athens, Greece, Oct. 2001.
6. Avrithis Y., Stamou G., Delopoulos A. and Kollias S., Intelligent Semantic Access to Audiovisual Content, *Lecture Notes in Artificial Intelligence*, Springer-Verlag, Vol. 2308, 2002, pp. 215-224.
7. Bowman C.M., Danzing P.B., Manber U. and Schwartz F., Scalable Internet resources discovery: research problems and approaches, *Communications of the ACM*, Vol. 37, pages 98-107, 1994.

8. Chan S.S.M., Qing L., Wu Y. and Zhuang Y., Accommodating hybrid retrieval in a comprehensive video database management system. *IEEE Trans. on Multimedia*, 4(2):146-159, June 2002.
9. Chen W. and Chang S.-F., VISMap: an interactive image/video retrieval system using visualization and concept maps. *In Proc. IEEE Int. Conf. on Image Processing*, volume 3, pages 588-591, 2001.
10. Ciocca G. and Schettini R., A relevance feedback mechanism for content – based image retrieval, *Information Processing and Management*, Vol. 35, pages 605-632, 1999.
11. Del Bimbo, A. *Visual Image Retrieval*, CA: Morgan Kaufmann, San Francisco, 1999.
12. Dubois D. and Prade H., The three semantics of fuzzy sets, *Fuzzy Sets and Systems*, Vol. 90,pages 142-150, 1997.
13. Zhao R. and W.I. Grosky, Narrowing the Semantic Gap-Improved Text-Based Web Document Retrieval Using Visual Features, *IEEE Trans. on Multimedia*, Special Issue on Multimedia Database, Vol. 4, No 2, June 2002.
14. Klir G. and Bo Yuan, *Fuzzy Sets and Fuzzy Logic, Theory and Applications*, New Jersey, Prentice Hall, 1995
15. Kraft D.H., Bordogna G., Passi G., Information Retrieval Systems: Where is the Fuzz?, *Proceedings of IEEE International Conference on Fuzzy Systems (FUZZ-IEEE)*, Anchorage, Alaska, May 1998
16. Maedche A., Motik B., Silva N. and Volz R., MAFRA – An Ontology MApping FRAmework in the Context of the SemanticWeb, *Proceedings of the Workshop on Ontology Transformation ECAI2002*, Lyon, France July 2002.
17. Miyamoto S., *Fuzzy Sets in Information Retrieval and Cluster Analysis*, Kluwer Academic Publishers, Dordrecht / Boston / London 1990.
18. Salembier P. and Smith J. R., MPEG-7 Multimedia description schemes, *IEEE Transactions on Circuits and Systems for Video Technology, 11(6):748-759*, Jun 2001.
19. Salton G. and McGill M.J., *Introduction to Modern Information Retrieval*, McGraw-Hill, New York, 1983.
20. Santini, S., *Exploratory Image Databases: Content-based Retrieval*, New York: Academic, 2001.
21. Theodoridis S. and Koutroumbas K., *Pattern Recognition*, Academic Press, 1998.
22. Tsechpenakis G., Akrivas G., Andreou G., Stamou G. and Kollias S., Knowledge – Assisted Video Analysis and Object Detection, *Proceedings of European Symposium on Intelligent Technologies, Hybrid Systems and their Implementation on Smart Adaptive Systems* Albufeira, Portugal, September 2002.
23. Wallace M., Athanasiadis T., Avrithis Y., Stamou G. and Kollias, S., A mediator system for hetero-lingual audiovisual content, *Proceedings of the International Conference on Multi-platform e-Publishing*, November 2004, Athens, Greece.
24. Wallace M., Avrithis Y., Stamou G. and Kollias S., Knowledge-based Multimedia Content Indexing and Retrieval, in Stamou G., Kollias S. (Editors), Multimedia Content and Semantic Web: Methods, Standards and Tools, Wiley, 2005.
25. Wallace M., Karpouzis K., Stamou G., Moschovitis G., Kollias S. and Schizas C., The Electronic Road: Personalised Content Browsing, *IEEE Multimedia 10(4)*, pp. 49-59, 2003

26. Yang Y., An Evaluation of Statistical Approaches to Text Categorization, *Journal of Information Retrieval*, Vol 1, No. 1/2, pp 67–88, 1999.
27. ISO/IEC JTC 1/SC 29 M4242, "Text of 15938-5 FDIS Information Technology – Multimedia Content Description Interface – Part 5 Multimedia Description Schemes," October 2001.

MIX
Papier aus verantwortungsvollen Quellen
Paper from responsible sources
FSC® C105338

If you have any concerns about our products,
you can contact us on
ProductSafety@springernature.com

In case Publisher is established outside the EU,
the EU authorized representative is:
Springer Nature Customer Service Center GmbH
Europaplatz 3, 69115 Heidelberg, Germany

Printed by Libri Plureos GmbH
in Hamburg, Germany